Shielded Metal Arc Welding

Shielded Metal Arc Welding

Edgar Gleeson

STATES
ACADEMIC PRESS
www.statesacademicpress.com

Shielded Metal Arc Welding
Edgar Gleeson
ISBN: 978-1-63989-811-4 (Hardback)

Published by States Academic Press,
109 South 5th Street,
Brooklyn, NY 11249, USA

Visit us on the World Wide Web at:
www.statesacademicpress.com

Cataloging-in-Publication Data

Shielded metal arc welding / Edgar Gleeson.
p. cm.
Includes bibliographical references and index.
ISBN 978-1-63989-811-4
1. Welding. 2. Welding--Equipment and supplies. 3. Welding--Safety measures.
I. Gleeson, Edgar.
TS227 .W45 2023
671.52--dc23

Table of Contents

Preface

Shielded metal arc welding (SMAW) is an arc welding process, which uses a consumable flux-covered electrode to lay the weld. The flux coating disintegrates as the weld is laid, emitting vapors in the process which serve as a shield and provide a layer of slag that protects the weld area from atmospheric contamination. SMAW employs minimal equipment such as a welding power supply, an electrode with a holder, a ground clamp and welding cables. The simplicity and versatility of this process makes it suitable for welding iron and steel, nickel, aluminium and copper alloys. It is used extensively in the repair and maintenance industry, industrial fabrication and for the construction of heavy steel structures. This book is compiled in such a manner, that it will provide in-depth knowledge about the theory and practice of shielded metal arc welding. It outlines the processes and applications of this welding process in detail. Coherent flow of topics, student-friendly language and extensive use of examples make this book an invaluable source of knowledge.

To facilitate a deeper understanding of the contents of this book a short introduction of every chapter is written below:

Chapter 1- Shielded metal arc welding is one of the most popular welding processes. It is a manual arc welding process. The electrode used is covered with a flux, and alternating or direct current from a power supply is used for laying the weld. This chapter is an introduction to shielded metal arc welding, and explores the processes of arc welding, shielded metal arc welding and gas shielded metal arc welding.

Chapter 2- Shielded metal arc welding process requires a constant power supply, an electrode and electrode holder, and welding cables. The topics covered in this chapter discuss the various types of equipment and the techniques of checking and choosing the best quality equipment for use in shielded metal arc welding.

Chapter 3- A welding joint is the point where two or more pieces of plastic or metal are joined together. It is of several types such as butt joint, corner joint, lap joint, edge joint and tee joint. This chapter examines such types of weld joints and the ways to perform these as well as the techniques of squaring, clamping and tacking joints and weld.

Chapter 4- Some of the problems associated with shielded metal arc welding are weld spatter, poor fusion, cracking, shallow distortion, etc. Besides the above, there can arise distortions, warping of thin plates and brittle welds. The strategies for the control and prevention of distortion, warping, corrosion, undercutting, spattering, etc. have been addressed in this chapter in great detail.

Chapter 5- Shielded metal arc welding (SMAW) is a dangerous process and requires strict adherence to proper precautions. Burns, shocks, arc eye along with exposure to UV radiation, gases and particulate matter are some of the common risks associated with this process. The aim of this chapter is to stress on the importance of safety while welding and highlights some of the risks associated with SMAW. It also suggests safety precautions for preventing such injuries.

Chapter 6- Science and technology has undergone rapid developments in the past few decades, which has resulted in innovative techniques of arc welding. Some of the variants of arc welding are flux-cored arc welding, gas metal arc welding, gas tungsten arc welding, plasma arc welding, submerged arc welding, etc. which have been discussed in great detail in this chapter.

I would like to share the credit of this book with my editorial team who worked tirelessly on this book. I owe the completion of this book to the never-ending support of my family, who supported me throughout the project.

Edgar Gleeson

CHAPTER 1

Introduction to Shielded Metal Arc Welding

Shielded metal arc welding is one of the most popular welding processes. It is a manual arc welding process. The electrode used is covered with a flux, and alternating or direct current from a power supply is used for laying the weld. This chapter is an introduction to shielded metal arc welding, and explores the processes of arc welding, shielded metal arc welding and gas shielded metal arc welding.

Arc Weld

Arc welding is one of several fusion processes for joining metals. By applying intense heat, metal at the joint between two parts is melted and caused to intermix - directly, or more commonly, with an intermediate molten filler metal. Upon cooling and solidification, a metallurgical bond is created. Since the joining is an intermixture of metals, the final weldment potentially has the same strength properties as the metal of the parts. This is in sharp contrast to non-fusion processes of joining (i.e. soldering, brazing etc.) in which the mechanical and physical properties of the base materials cannot be duplicated at the joint.

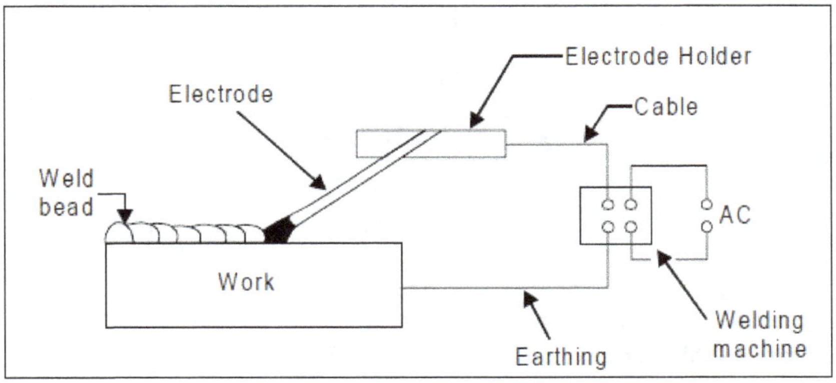

The basic arc-welding circuit

In arc welding, the intense heat needed to melt metal is produced by an electric arc. The arc is formed between the actual work and an electrode (stick or wire) that is manually or mechanically guided along the joint. The electrode can either be a rod with the purpose of simply carrying the current between the tip and the work. Or, it may be a specially prepared rod or wire that not only conducts the current but also melts and supplies filler metal to the joint. Most welding in the manufacture of steel products uses the second type of electrode.

Basic Welding Circuit

The basic arc-welding circuit is illustrated in figure above. An AC or DC power source, fitted with whatever controls may be needed, is connected by a work cable to the workpiece and by a "hot" cable to an electrode holder of some type, which makes an electrical contact with the welding electrode.

An arc is created across the gap when the energized circuit and the electrode tip touches the workpiece and is withdrawn, yet still with in close contact.

The arc produces a temperature of about 6500°F at the tip. This heat melts both the base metal and the electrode, producing a pool of molten metal sometimes called a "crater." The crater solidifies behind the electrode as it is moved along the joint. The result is a fusion bond.

Arc Shielding

However, joining metals requires more than moving an electrode along a joint. Metals at high temperatures tend to react chemically with elements in the air - oxygen and nitrogen. When metal in the molten pool comes into contact with air, oxides and nitrides form which destroy the strength and toughness of the weld joint. Therefore, many arc-welding processes provide some means of covering the arc and the molten pool with a protective shield of gas, vapor, or slag. This is called arc shielding. This shielding prevents or minimizes contact of the molten metal with air. Shielding also may improve the weld. An example is a granular flux, which actually adds deoxidizers to the weld.

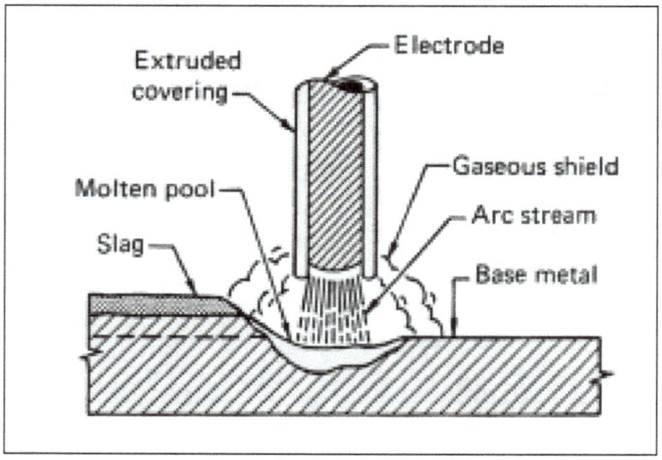

This shows how the coating on a coated (stick) electrode provides a gaseous shield around the arc and a slag covering on the hot weld deposit.

Figure above illustrates the shielding of the welding arc and molten pool with a Stick electrode. The extruded covering on the filler metal rod, provides a shielding gas at the point of contact while the slag protects the fresh weld from the air.

The arc itself is a very complex phenomenon. In-depth understanding of the physics of the arc is of little value to the welder, but some knowledge of its general characteristics can be useful.

Nature of the Arc

An arc is an electric current flowing between two electrodes through an ionized column of gas. A negatively charged cathode and a positively charged anode create the intense heat of the welding arc. Negative and positive ions are bounced off of each other in the plasma column at an accelerated rate.

In welding, the arc not only provides the heat needed to melt the electrode and the base metal, but under certain conditions must also supply the means to transport the molten metal from the tip of the electrode to the work. Several mechanisms for metal transfer exist. Two (of many) examples include:

1. Surface Tension Transfer - a drop of molten metal touches the molten metal pool and is drawn into it by surface tension
2. Spray Arc - the drop is ejected from the molten metal at the electrode tip by an electric pinch propelling it to the molten pool (great for overhead welding)

If an electrode is consumable, the tip melts under the heat of the arc and molten droplets are detached and transported to the work through the arc column. Any arc welding system in which the electrode is melted off to become part of the weld is described as metal-arc. In carbon or tungsten (TIG) welding there are no molten droplets to be forced across the gap and onto the work. Filler metal is melted into the joint from a separate rod or wire.

More of the heat developed by the arc is transferred to the weld pool with consumable electrodes. This produces higher thermal efficiencies and narrower heat-affected zones.

Since there must be an ionized path to conduct electricity across a gap, the mere switching on of the welding current with an electrically cold electrode posed over it will not start the arc. The arc must be ignited. This is caused by either supplying an initial voltage high enough to cause a discharge or by touching the electrode to the work and then withdrawing it as the contact area becomes heated.

Arc welding may be done with direct current (DC) with the electrode either positive or negative or alternating current (AC). The choice of current and polarity depends on the process, the type of electrode, the arc atmosphere, and the metal being welded.

Different Types of Arc Welding

There are different forms of arc welding, such as MIG or stick welding.

Flux-cored Arc Welding (FCAW)

Flux-cored arc welding uses tubular electrodes that are filled with flux. It's much less brittle than the coatings on SMAW electrodes and preserves most of the alloying benefits. The emissive fluxes shield the welding arc from the air, or shielding gases might be used if nonemissive fluxes are required. It's popular when welding heavy sections an inch or more thick thanks to the higher weld-metal deposition rate.

Gas Metal Arc Welding (GMAW)

Gas metal arc welding, also known as MIG welding, shields the welding arc with a gas such as argon or helium or even a mixture. Deoxidizers in the electrodes can prevent oxidation which makes it possible to weld mulitple layers. It's a simple, versatile, and economical welding process. The temperatures are also relatively low and it is used for thin sheet and sections. It can easily be automated.

Gas Tungsten Arc Welding (GTAW)

Gas tungsten arc welding is also known as TIG welding. It uses tungsten electrodes as one pole of the arc in order to create the required heat. The gas is argon, helium, or a mixture of those two. Filler wires provide the molten material if it is necessary. This process is good for thin materials and the filler wires are similar in composition to whatever is being welded.

Plasma Arc Welding (PAW)

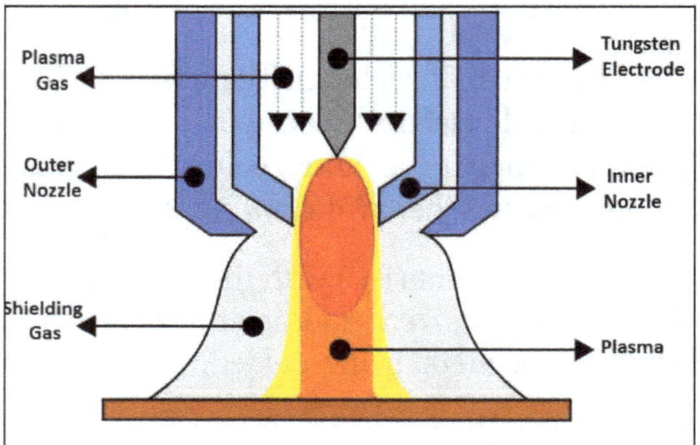

Principle of Plasma Arc Welding

Plasma arc welding has ionized gases and electrodes that generate hot plasma jets that are aimed at the welding area. These jets are extremely hot. The concentration of higher energy is good for narrower and deeper welds as well as an increase in welding speeds.

Shielded Metal Arc Welding (SMAW)

Shielded metal arc welding is one of the simplest, oldest, and most versatile welding methods. The arc comes from a coated electrode tip being touched to the workpiece and then withdrawn to maintain the arc. The heat that is generated melts the tip, coating, and base metal and the weld is formed out of that alloy when it solidifies. Slag that is formed and protects the weld from oxides, inclusions, and nitrides has to be removed after every pass. This is commonly used in pipeline work, shipbuilding, and construction.

Submerged Arc Welding (SAW)

Submerged arc welding has a granular flux that is fed into the weld zone that forms a thick layer, completely covering the molten zone and preventing sparks and spatter. It allows for deeper heat penetration since it acts like a thermal insulator. The process is limited to horizontal welds and used for high speed sheet or plate steel welding. It can be semiautomatic or automatic. The flux can be recovered and treated then used again. This method provides 4-10 times as much productivity as shielded metal arc welding.

How to do Arc Weld Properly

Arc welding is the most elementary type of welding which is not only minimalist in terms of the paraphernalia required but also among the least expensive procedures to start out with.

If you are a DIY enthusiast looking to know your way about the arc welding procedure, here is a step-by-step guide for you to refer to:

1. Getting the Materials and Welding Area Ready

Before setting out to weld, it is important that you get all the necessary welding paraphernalia ready. Select your welding electrode, welding machine, electrode holder, workpiece, ground clamp et al. The area you want to perform the welding operation in should also be checked for the presence of any flammable substances which, in case, found, should be removed. In case you do not have an area to dedicate your welding activities to, have welding curtains installed around your welding zone so as to prevent people in proximity from the direct exposure of UV rays and glares.

2. Using Protective Gear

When it comes to a job such as welding, if you do not stick by the rule book and give even the slightest of seemingly trivial precautions or rules a miss, you'd end up doing yourself more harm than good. Before you set out to perform your first weld, ensure you have thoroughly gotten hold of each and every safety wear for yourself—from safety glasses to welding gloves. The process of welding gives off sparks and spatters which run the risk of flying straight into your eye. Also seeing welding sparks directly with the naked eye can damage visibility. Thus, you must wear welding glasses at any cost. Sparks from welding could not only ignite a fire but could also cause burns to your skin. For this reason, you must always remember to wear welding gloves as well.

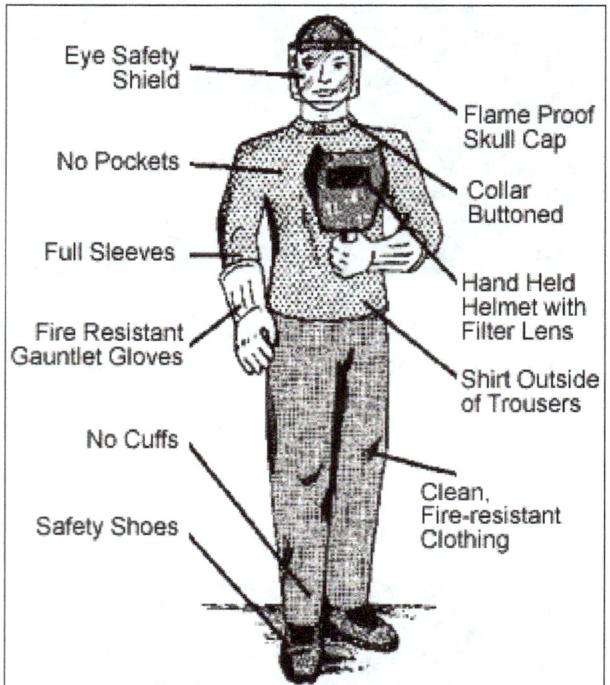

Select clothing to provide maximum protection from sparks and hot metals

3. Making Adjustments in the Welding Machine

A welding machine usually operates at an amperage ranging from anywhere between 90 – 120 amps. However, you could always tune the amperage depending on the material that you intend to weld.

4. Cleaning the Metal Prior to Welding

You must make it a point to clean the metal to be welded with the help of either a wire brush or a grinder. The process helps clear the workpiece of any kind of rust accumulation or old paint remains. You could also use acetone to clear the metal off of ant oil deposits. You must not resort to the use of any chlorinated solvent to clean the workpiece; upon welding, it could trigger a lethal reaction.

5. Securing the Joint and Striking the Arc

Secure the workpiece firmly in place with the help of a clamp and vise. Once done, you could strike the arc by gently touching and tugging the metal using the tip of the electrode. The process has to be repeated on and off a few times.

6. Building a Weld Pool

Once you have established the arc, the electrode has to be maintained at a fixed distance of about one-eighth of an inch from the metal so as to allow it to melt and form a weld pool.

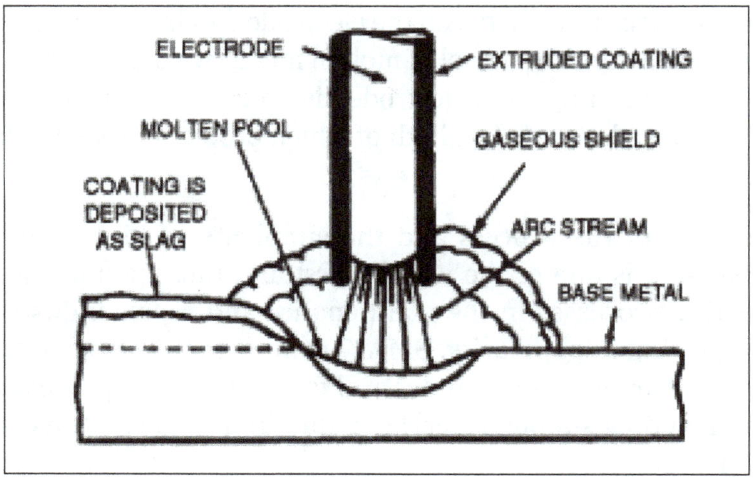

7. Finishing the Weld

Once you have finished the joining process, you can pull the electrode back, allow the workpiece to cool and clear the excess slag with the help of a chipping hammer.

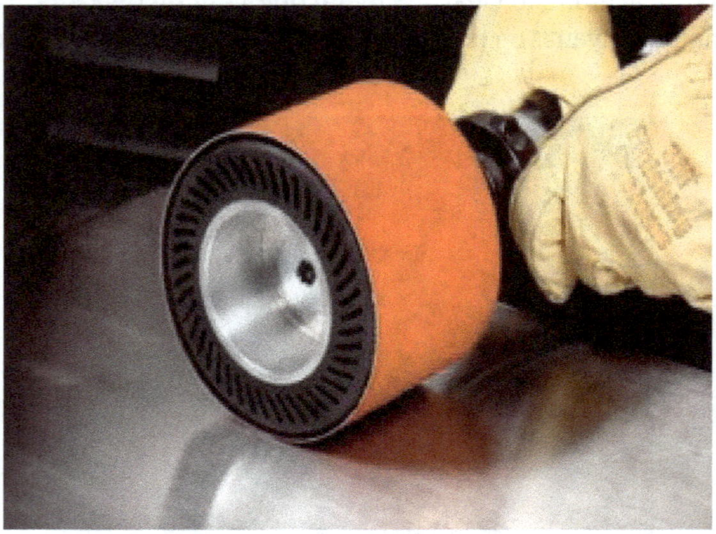

Practise the aforementioned steps, over and over again, under the supervision of an experienced welder. Also, there is more to why you should start with arc welding of all the different types of welding procedures that exist; this is because upon gaining good practice, you could improvise your welds to include the multitude of specialised electrodes available.

Shielded Metal Arc Welding

Shielded metal arc welding (SMAW), also known as manual metal arc welding (MMA or MMAW), flux shielded arc welding or informally as stick welding, is a manual arc welding process that uses a consumable electrode covered with a flux to lay the weld.

An electric current, in the form of either alternating current or direct current from a welding power supply, is used to form an electric arc between the electrode and the metals to be joined. The workpiece and the electrode melts forming a pool of molten metal (weld pool) that cools to form a joint. As the weld is laid, the flux coating of the electrode disintegrates, giving off vapors that serve as a shielding gas and providing a layer of slag, both of which protect the weld area from atmospheric contamination.

Because of the versatility of the process and the simplicity of its equipment and operation, shielded metal arc welding is one of the world's first and most popular welding processes. It dominates other welding processes in the maintenance and repair industry, and though flux-cored arc welding is growing in popularity, SMAW continues to be used extensively in the construction of heavy steel structures and in industrial fabrication. The process is used primarily to weld iron and steels (including stainless steel) but aluminium, nickel and copper alloys can also be welded with this method.

Operation

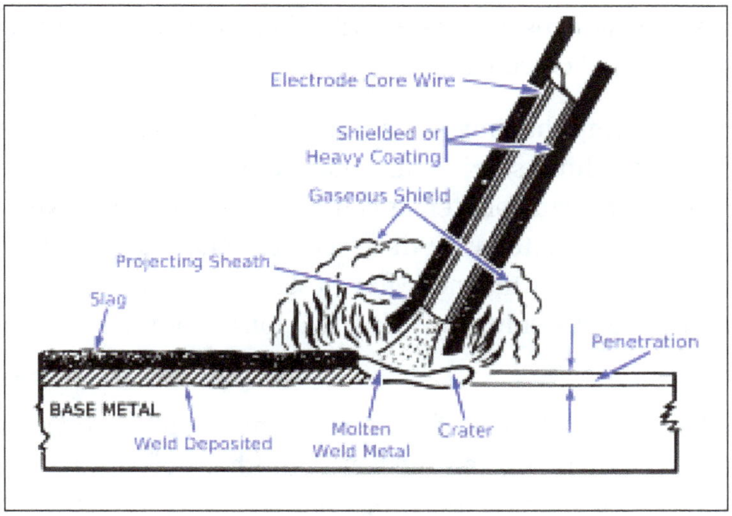

SMAW weld area

To strike the electric arc, the electrode is brought into contact with the workpiece by a very light touch of the electrode to the base metal. The electrode is then pulled back slightly. This initiates the arc and thus the melting of the workpiece and the consumable electrode, and causes droplets of the electrode to be passed from the electrode to the weld pool. Striking an arc, which varies widely based upon electrode and workpiece composition, can be the hardest skill for beginners. The orientation of the electrode to workpiece is where most stumble, if the electrode is held at a perpendicular angle to the workpiece the tip will likely stick to the metal which will fuse the electrode to the workpiece which will cause it to heat up very rapidly. The tip of the electrode needs to be at a lower angle to the workpiece, which allows the weld pool to flow out of the arc. As the electrode melts, the flux covering disintegrates, giving off shielding gases that protect the weld area from oxygen and other atmospheric gases. In addition, the flux provides molten slag which covers the filler metal as it travels from the electrode to the weld pool. Once part of the weld pool, the slag floats to the surface and protects the weld from contamination as it solidifies. Once hardened, it must be chipped away to reveal the finished weld. As welding progresses and the electrode melts, the welder must periodically stop welding to remove the remaining electrode stub and insert a new electrode into the electrode holder. This activity, combined with chipping away the slag, reduces the amount of time that the welder can spend laying the weld, making SMAW one of the least efficient welding processes. In general, the operator factor, or the percentage of operator's time spent laying weld, is approximately 25%.

The actual welding technique utilized depends on the electrode, the composition of the workpiece, and the position of the joint being welded. The choice of electrode and welding position also determine the welding speed. Flat welds require the least operator skill, and can be done with electrodes that melt quickly but solidify slowly. This permits higher welding speeds.

Sloped, vertical or upside-down welding requires more operator skill, and often necessitates the use of an electrode that solidifies quickly to prevent the molten metal from flowing out of the weld pool. However, this generally means that the electrode melts less quickly, thus increasing the time required to lay the weld.

Quality

The most common quality problems associated with SMAW include weld spatter, porosity, poor fusion, shallow penetration, and cracking.

Weld spatter, while not affecting the integrity of the weld, damages its appearance and increases cleaning costs. Secondary finishing services are often required due to the aesthetic appearance caused by the occurrence of molten splatter. It can be caused by excessively high current, a long arc, or arc blow, a condition associated with direct current characterized by the electric arc being deflected away from the weld pool by magnetic forces. Arc blow can also cause porosity in the weld, as can joint contamination, high welding speed, and a long welding arc, especially when low-hydrogen electrodes are used.

Porosity, often not visible without the use of advanced nondestructive testing methods, is a serious concern because it can potentially weaken the weld. Another defect affecting the strength of the weld is poor fusion, though it is often easily visible. It is caused by low current, contaminated joint surfaces, or the use of an improper electrode.

Shallow penetration, another detriment to weld strength, can be addressed by decreasing welding speed, increasing the current or using a smaller electrode. Any of these weld-strength-related defects can make the weld prone to cracking, but other factors are involved as well. High carbon, alloy or sulfur content in the base material can lead to cracking, especially if low-hydrogen electrodes and preheating are not employed. Furthermore, the workpieces should not be excessively restrained, as this introduces residual stresses into the weld and can cause cracking as the weld cools and contracts.

Safety

SMAW welding, like other welding methods, can be a dangerous and unhealthy practice if proper precautions are not taken. The process uses an open electric arc, which presents a risk of burns which are prevented by personal protective equipment in the form of heavy leather gloves and long sleeve jackets. Additionally, the brightness of the weld area can lead to a condition called arc eye, in which ultraviolet light causes inflammation of the cornea and can burn the retinas of the eyes. Welding helmets with dark face plates are worn to prevent this exposure, and in recent years, new helmet models have been produced that feature a face plate that self-darkens upon exposure to high amounts of UV light. To protect bystanders, especially in industrial environments, translucent welding curtains often surround the welding area. These curtains, made of a polyvinyl chloride plastic film, shield nearby workers from exposure to the UV light from the electric arc, but should not be used to replace the filter glass used in helmets.

In addition, the vaporizing metal and flux materials expose welders to dangerous gases and particulate matter. The smoke produced contains particles of various types of oxides. The size of the particles in question tends to influence the toxicity of the fumes, with smaller particles presenting a greater danger. Additionally, gases like carbon dioxide and ozone can form, which can prove dangerous if ventilation is inadequate. Some of the latest welding masks are fitted with an electric powered fan to help disperse harmful fumes.

Application and Materials

Shielded metal arc welding is one of the world's most popular welding processes, accounting for over half of all welding in some countries. Because of its versatility and simplicity, it is particularly dominant in the maintenance and repair industry, and is heavily used in the construction of steel structures and in industrial fabrication. In recent years its use has declined as flux-cored arc welding has expanded in the construction industry and gas metal arc welding has become more popular in industrial environments. However, because of the low equipment cost and wide applicability, the process will likely remain popular, especially among amateurs and small businesses where specialized welding processes are uneconomical and unnecessary.

SMAW is often used to weld carbon steel, low and high alloy steel, stainless steel, cast iron, and ductile iron. While less popular for nonferrous materials, it can be used on nickel and copper and their alloys and, in rare cases, on aluminium. The thickness of the material being welded is bounded on the low end primarily by the skill of the welder, but rarely does it drop below 1.5 mm (0.06 in). No upper bound exists: with proper joint preparation and use of multiple passes, materials of virtually unlimited thicknesses can be joined. Furthermore, depending on the electrode used and the skill of the welder, SMAW can be used in any position.

Equipment

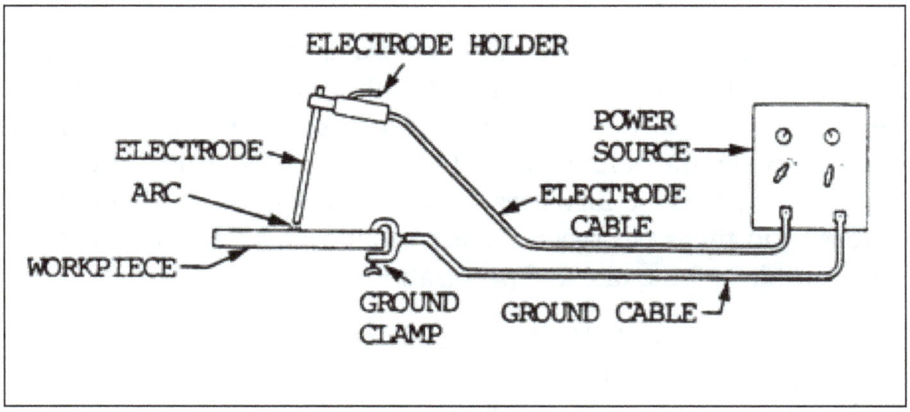

SMAW system setup

Shielded metal arc welding equipment typically consists of a constant current welding power supply and an electrode, with an electrode holder, a 'ground' clamp, and welding cables (also known as welding leads) connecting the two.

Power Supply

The power supply used in SMAW has constant current output, ensuring that the current (and thus the heat) remains relatively constant, even if the arc distance and voltage change. This is important because most applications of SMAW are manual, requiring that an operator hold the torch. Maintaining a suitably steady arc distance is difficult if a constant voltage power source is used instead, since it can cause dramatic heat variations and make welding more difficult. However, because the current is not maintained absolutely constant, skilled welders performing complicated welds can vary the arc length to cause minor fluctuations in the current.

A high output welding power supply for SMAW, GTAW, MIG, Flux-Cored, & Gouging

The preferred polarity of the SMAW system depends primarily upon the electrode being used and the desired properties of the weld. Direct current with a negatively charged electrode (DCEN) causes heat to build up on the electrode, increasing the electrode melting rate and decreasing the depth of the weld. Reversing the polarity so that the electrode is positively charged (DCEP) and the workpiece is negatively charged increases the weld penetration. With alternating current the polarity changes over 100 times per second, creating an even heat distribution and providing a balance between electrode melting rate and penetration.

Typically, the equipment used for SMAW consists of a step-down transformer and for direct current models a rectifier, which converts alternating current into direct current. Because the power normally supplied to the welding machine is high-voltage alternating current, the welding transformer is used to reduce the voltage and increase the current. As a result, instead of 220 V at 50 A, for example, the power supplied by the transformer is around 17–45 V at currents up to 600 A. A number of different types of transformers can be used to produce this effect, including multiple coil and inverter machines, with each using a different method to manipulate the welding current. The multiple coil type adjusts the current by either varying the number of turns in the coil (in tap-type transformers) or by varying the distance between the primary and secondary coils (in movable coil or movable core transformers). Inverters, which are smaller and thus more portable, use electronic components to change the current characteristics.

Electrical generators and alternators are frequently used as portable welding power supplies, but because of lower efficiency and greater costs, they are less frequently used in industry. Maintenance also tends to be more difficult, because of the complexities of using a combustion engine as a power source. However, in one sense they are simpler: the use of a separate rectifier is unnecessary because they can provide either AC or DC. However, the engine driven units are most practical in field work where the welding often must be done out of doors and in locations where transformer type welders are not usable because there is no power source available to be transformed.

In some units the alternator is essentially the same as that used in portable generating sets used to supply mains power, modified to produce a higher current at a lower voltage but still at the 50 or 60 Hz grid frequency. In higher-quality units an alternator with more poles is used and supplies current at a higher frequency, such as 400 Hz. The smaller amount of time the high-frequency waveform spends near zero makes it much easier to strike and maintain a stable arc than with the cheaper grid-frequency sets or grid-frequency mains-powered units.

Electrode

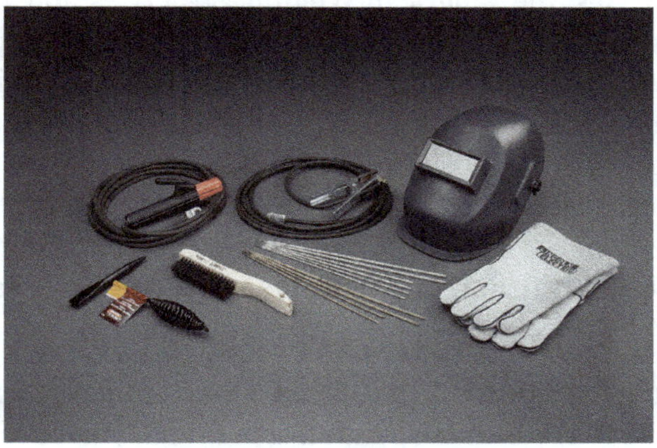

Various accessories for SMAW

The choice of electrode for SMAW depends on a number of factors, including the weld material, welding position and the desired weld properties. The electrode is coated in a metal mixture called flux, which gives off gases as it decomposes to prevent weld contamination, introduces deoxidizers to purify the weld, causes weld-protecting slag to form, improves the arc stability, and provides alloying elements to improve the weld quality. Electrodes can be divided into three groups—those designed to melt quickly are called "fast-fill" electrodes, those designed to solidify quickly are called "fast-freeze" electrodes, and intermediate electrodes go by the name "fill-freeze" or "fast-follow" electrodes. Fast-fill electrodes are designed to melt quickly so that the welding speed can be maximized, while fast-freeze electrodes supply filler metal that solidifies quickly, making welding in a variety of positions possible by preventing the weld pool from shifting significantly before solidifying.

The composition of the electrode core is generally similar and sometimes identical to that of the base material. But even though a number of feasible options exist, a slight difference in alloy composition can strongly impact the properties of the resulting weld. This is especially true of alloy steels such as HSLA steels. Likewise, electrodes of compositions similar to those of the base materials are often used for welding nonferrous materials like aluminium and copper. However, sometimes it is desirable to use electrodes with core materials significantly different from the base material. For example, stainless steel electrodes are sometimes used to weld two pieces of carbon steel, and are often utilized to weld stainless steel workpieces with carbon steel workpieces.

Electrode coatings can consist of a number of different compounds, including rutile, calcium fluoride, cellulose, and iron powder. Rutile electrodes, coated with 25%–45% TiO_2, are characterized by ease of use and good appearance of the resulting weld. However, they create welds with high hydrogen content, encouraging embrittlement and cracking. Electrodes containing calcium fluoride (CaF_2), sometimes known as basic or low-hydrogen electrodes, are hygroscopic and must be stored in dry conditions. They produce strong welds, but with a coarse and convex-shaped joint surface. Electrodes coated with cellulose, especially when combined with rutile, provide deep weld penetration, but because of their high moisture content, special procedures must be used to prevent excessive risk of cracking. Finally, iron powder is a common coating additive that increases the rate at which the electrode fills the weld joint, up to twice as fast.

To identify different electrodes, the American Welding Society established a system that assigns electrodes with a four- or five-digit number. Covered electrodes made of mild or low alloy steel carry the prefix *E*, followed by their number. The first two or three digits of the number specify the tensile strength of the weld metal, in thousand pounds per square inch (ksi). The penultimate digit generally identifies the welding positions permissible with the electrode, typically using the values 1 (normally fast-freeze electrodes, implying all position welding) and 2 (normally fast-fill electrodes, implying horizontal welding only). The welding current and type of electrode covering are specified by the last two digits together. When applicable, a suffix is used to denote the alloying element being contributed by the electrode.

Common electrodes include the E6010, a fast-freeze, all-position electrode with a minimum tensile strength of 60 ksi (410 MPa) which is operated using DCEP, and provides deep weld penetration with a forceful arc capable of burning through light rust or oxides on the workpiece. E6011 is similar except its flux coating allows it to be used with alternating current in addition to DCEP. E7024 is a fast-fill electrode, used primarily to make flat or horizontal welds using AC, DCEN, or DCEP. Examples of fill-freeze electrodes are the E6012, E6013, and E7014, all of which provide a compromise between fast welding speeds and all-position welding.

Process Variations

Though SMAW is almost exclusively a manual arc welding process, one notable process variation exists, known as gravity welding or gravity arc welding. It serves as an automated version of the traditional shielded metal arc welding process, employing an electrode holder attached to an inclined bar along the length of the weld. Once started, the process continues until the electrode is spent, allowing the operator to manage multiple gravity welding systems. The electrodes employed (often E6027 or E7024) are coated heavily in flux, and are typically 71 cm (28 in) in length and about 6.35 mm (0.25 in) thick. As in manual SMAW, a constant current welding power supply is used, with either negative polarity direct current or alternating current. Due to a rise in the use of semiautomatic welding processes such as flux-cored arc welding, the popularity of gravity welding has fallen as its economic advantage over such methods is often minimal. Other SMAW-related methods that are even less frequently used include firecracker welding, an automatic method for making butt and fillet welds, and massive electrode welding, a process for welding large components or structures that can deposit up to 27 kg (60 lb) of weld metal per hour.

How to do Shielded Metal Arc Welding

Step 1. Pre-welding Familiarization

Prior to welding, you should read the user-manual of your SMAW welding machine. Welders come with a wide variety of features, and can utilize AC current, DC current, and vary on amperage. Select an appropriate rod for your machine. A 6013 rod will work for most machines, and an amperage setting of approximately 100 amps using AC current will work best with the rod. Again, the user manual provided with the welding machine will layout the procedure for checking this configuration.

Step 2. Safety

Ensure you are in a well ventilated area before beginning welding. SMAW arc welding produces toxic fumes. Ensure you are wearing proper protective equipments. Proper equipment includes a long sleeve shirt, long pants, leather shoes, and gauntlet gloves. The long sleeve shirt is required because arc rays can cause minor skin burns identical to sun burns. Gauntlet Gloves are required because they protect your hands from the immense heat created during the welding process. The gloves must be loose fitting so that in case they catch on fire, they can be quickly and easily separated from your hands. Lastly you will need a face shield or helmet with a #10 shaded lens dawned or ready.

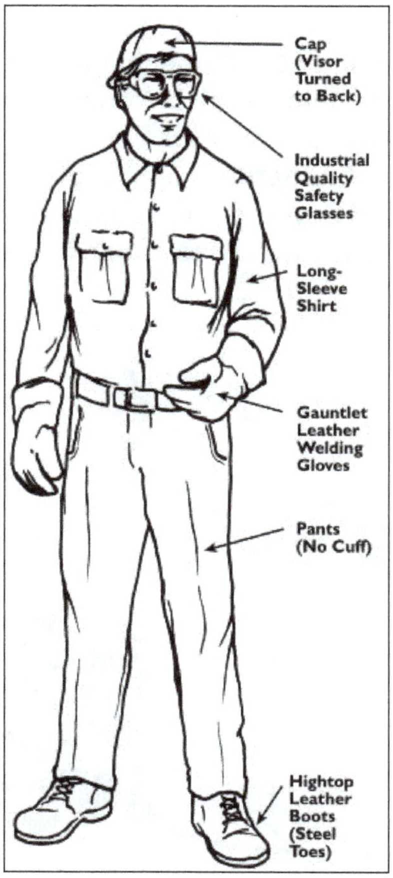

Step 3. Setup

Ensure the welding equipment is properly set up, plugged in and the ground is connected to the piece of metal you are working on. SMAW welding uses high voltage electricity which runs from the outlet through the machine, into the stick, through the arc, into the piece of metal being welded, through the ground clamp, back into the machine, and back into the outlet. This flow of electricity is critical to the weld, and compromising it could result in sever shock of the welder. Note: The picture below depicts the welding ground attached to the table. This is done if a metal welding table is used. For other projects, the clamp should be directly attached to the piece of metal being worked on.

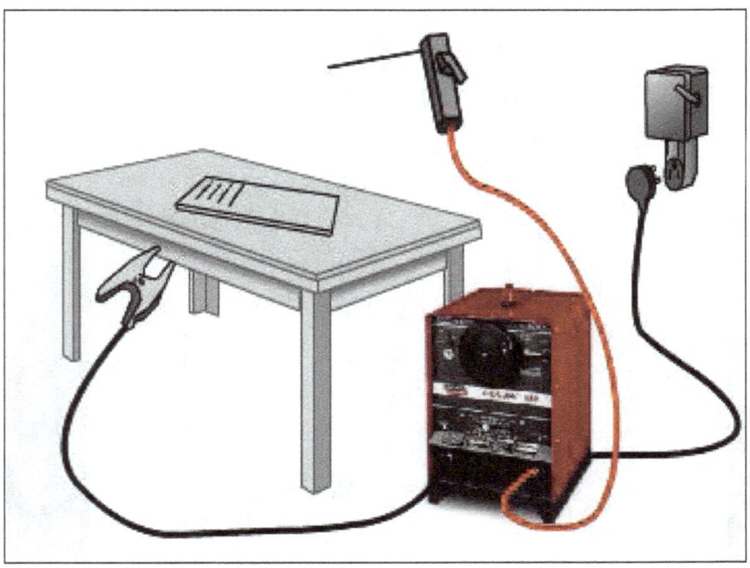

Keep the pliers, slag hammer, wire brush, and bucket of water near the area you are working. You will use these tools directly after laying your bead on the metal.

Step 4. Strike an Arc and Position Rod

With the piece of metal prepped and ready, it is time to begin welding. Turn the SMAW welding machine on, and down your helmet. Strike an arc by quickly brushing the rod against the piece of metal that you are attempting to weld. After brushing the rod across the metal, try to maintain a distance of about 1/16" between the end of the rod, and the piece of metal you are welding. This distance is crucial to make a good weld. You will notice a bright arc that is constant and stable if done correctly. With the arc struck, the rod should be held pointing directly away from you and in line with the intended weld. The rod should be held at a 45 degree vertical angle with the metal being welded.

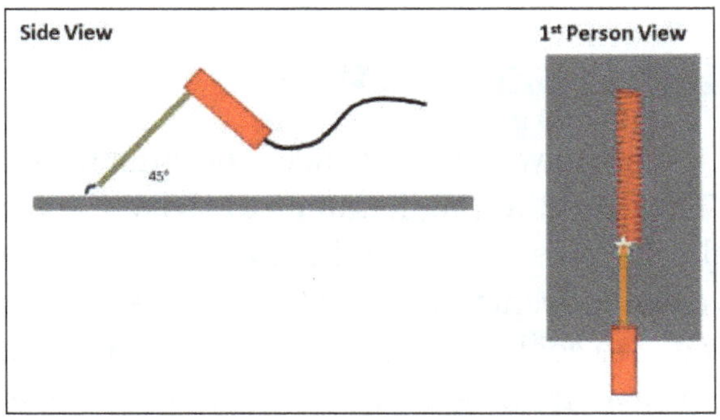

Note: The picture above depicts the motion used only, and not the molten metal, or "pool" produced when welding. Using the motion described below will produce a pool that is overlapped with each pass so that no space can be seen between zig-zag strokes.

Step 5. Begin Welding the Bead

Now, with the arc struck, and the rod position correct, begin welding by pulling the rod towards you in a zig-zag or semi-circular motion. The proper angle and the 1/16" distance between the rod and the metal should be maintained throughout the entire welding process. Remember, that as long as the arc is struck, the rod will continue to get shorter because metal from the rod is "flowing" across the arc and onto the metal being welded. Therefore it is difficult to maintain the 1/16" arc distance and will require lots of practice.

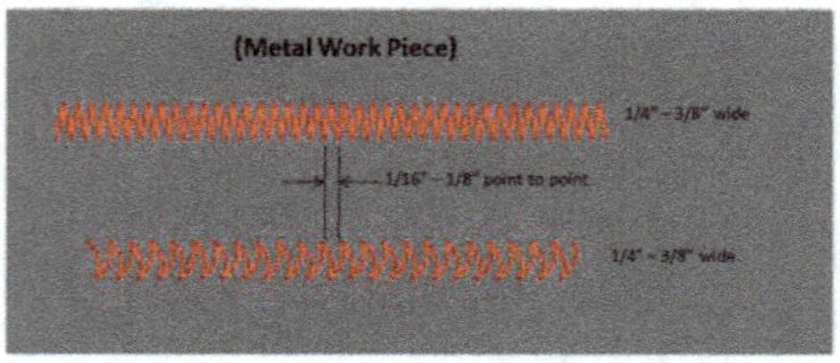

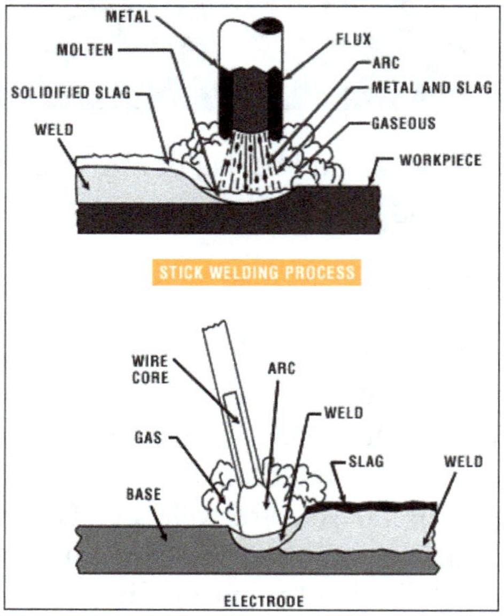

Advantage Fabricated Metals, Inc.

Note: Each stroke of the zig-zag motion should overlap the existing pool by about 50%.

Step 6. Remove Slag and Examine the Weld

After the weld is complete, the metal will be extremely hot. First ensure you turned the power off on the SMAW arc welder. Then use the slag hammer to strike the bead you just made. A small amount of crusted material called the slag should chip away from the weld revealing the actual bead. Use the wire brush to clean the remaining slag from the bead. Gripping the metal with pliers, dunk it into the bucket of water to cool the metal. After removing from the water, wipe the metal dry clean with the wire brush again if needed. Finally examine your weld.

The welding speed will vary with different welding rods, however, the correct speed should be fairly slow and extremely steady. If done correctly the weld should look like it does below. If the weld is not continuos some common faults are motions that are spread too far apart, or moving too fast across the metal. It will take several attempts before you are able to successfully lay down a bead weld.

How to Avoid Troubleshooting in Shielded Metal Arc Welding

By following a few simple tips, even beginners can learn how to spot common weld defects and fix them to create a high-quality weld.

1. Select "SMAW-Friendly" Steel

Whenever possible, select AISI-SAE 1015 to 1025 steels with 0.1 percent maximum silicon and sulfur content under 0.035 percent. Using these steels will make the SMAW process easier because they can be welded at fast speeds with minimum cracking tendencies.

If you are welding with low-alloy steels and carbon steels with chemistry compositions higher than this normal range, they will have a greater tendency to crack, particularly when welding on heavy plate and rigid structures. In addition, steels with high sulfur and phosphorus contents are not recommended for production welding. If they must be welded, use small-diameter, low-hydrogen electrodes. A slow travel speed will further keep the puddle molten, allowing gas bubbles time to boil out, creating a better finished weld.

2. Match the Joint Position and Electrode with the Metal

Joint position can greatly affect finished weld quality. When welding on 10- to 18-gauge sheet steel, the fastest travel speeds are obtained with the work positioned at 45 to 75 degrees downhill. Also, don't overweld or make a weld that is larger than necessary for the sake of joint strength—this may lead to burn-through.

When welding mild steel plate that is 3/16 in. thick or more, it is best to position the work flat, as this will make it easier for you to manipulate the electrode. Last, it is best to weld high-carbon and low-alloy steel plate in the level position.

3. Follow the Principles for Joint Geometry and Fit-up

Joint dimensions are chosen with fast welding speed and good weld quality in mind. Proper joint geometry is based on a few simple principles. First, fit-up must be consistent for the entire joint. Since sheet metal and most fillet and lap joints are tightly clamped for their entire length, gaps or bevels must be controlled accurately over the entire joint. Any variations in a joint will force you to slow your welding speed to avoid burn-through and manipulate the electrode to adjust for the fit-up variation.

Second, you need a bevel that will aid in good bead shape and penetration. An insufficient bevel prevents the electrode from getting into the joint. For example, a deep, narrow bead with an insufficient bevel may lack penetration, making it prone to cracking.

Third, a root opening is needed for full penetration. The root opening must be consistent with the diameter of the electrode being used. An excessive root opening wastes weld metal and slows welding speed.

And last, a root face or a backup strip is required for fast welding and good quality. Feather-edge preparations require a slow, costly seal bead. However, double-V butt joints without a land are practical when the seal bead cost is offset by easier edge preparation and the root opening can be limited to approximately 3/32 in.

In general, weld seal beads on flat work with a 3/16-in. E6010 at approximately 150 amps direct current electrode positive (DCEP). Use 1/8 in. at approximately 90 amps DCEP for vertical, overhead, and horizontal butt welds. For low-hydrogen and seal beads, weld with an EXX18 electrode at approximately 170 amps.

4. Avoid Buildup and Overwelding

Fillets should have equal legs, and the bead surface should be nearly flat. Buildup shouldn't exceed 1/16 in. Extra buildup is costly in material and time, adds little to weld strength, and increases distortion. For example, doubling the size of a fillet requires four times as much weld metal.

5. Clean the Joint Before you Weld

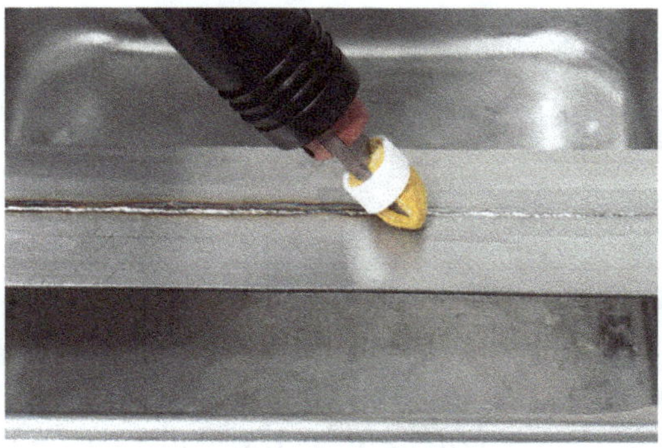

To avoid porosity and attain the ideal weld travel speed, it is important to remove excessive scale, rust, moisture, paint, oil, and grease from the surface of joints. If such elements cannot be removed, use E6010 (5P+) or E6011 (35 or 180) electrodes to penetrate the contaminants and deeply into the base metal. Slow the travel speed to allow time for gas bubbles to boil out of the molten weld before it freezes.

6. Choose the Right Electrode Size

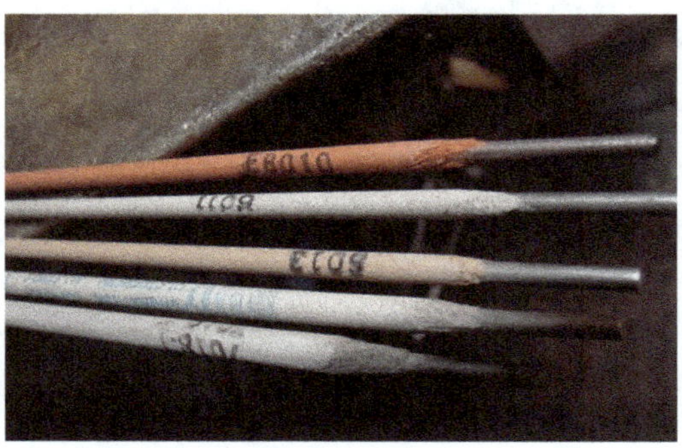

Large electrodes are meant for welding at high currents for high deposition rates. Therefore, use the largest electrode that is practical for your application and consistent with good weld quality.

Electrode size sometimes may be limited, especially on sheet metal and root passes where burn-through can occur. As a rule, 3/16 in. is the maximum electrode size practical for vertical and overhead welding, while 5/32 in. is the maximum size for low-hydrogen welding. In addition, joint dimensions sometimes limit the electrode diameter that will fit into the joint.

Gas Shielded Metal Arc Welding

The gas shielded metal arc welding (GSMAW) process configuration consists of several components and consumables are arranged in below figure. The main components and consumables are:

(1) Welding power source

(2) Remote controller

(3) Wire feeder

(4) Welding torch

(5) Shielding gas cylinder and regulator

(6) Welding wire

In addition, a water circulator for a water-cooled welding torch. Below figure shows an overall view of power source, wire feeder and welding torch. With the GSMAW process using a constant-voltage power source, the electrodes wire is fed at a constant speed that matches the welding current while the arc length is remained almost constant by the self – correction conduit tube and welding torch. During welding, the arc and molten pool are shielded with a shielding gas to prevent them from the adverse effects of nitrogen and oxygen in the atmosphere. The type of welding wire chooses the proper kind of shielding gas to ensure intended usability and weldability; however, the wires for mild steel and high strength steel mostly use CO_2 and 75-80% Ar/bal.CO_2 mixtures for general applications.

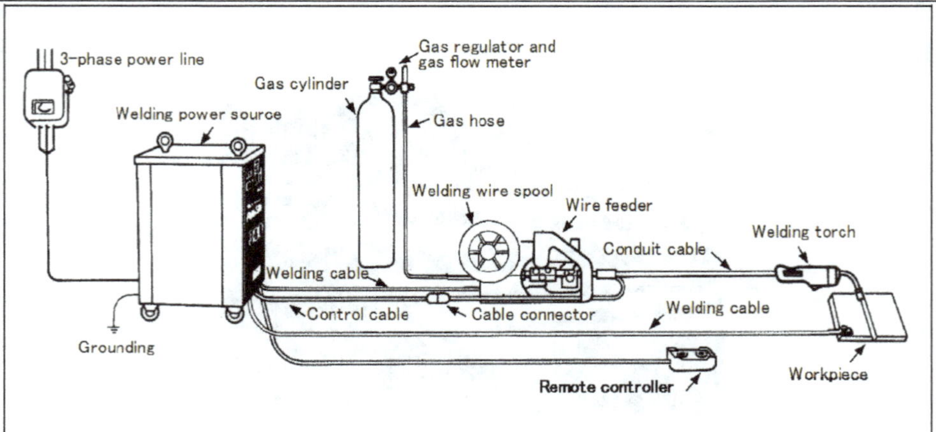

Typical arrangement of the gas shielded metal arc welding process

Process for GSMAW

After it is confirmed that electrical connection and shielding gas passage connection (and water – cooling circuit connection for a water- cooled welding torch) are correctly completed, follow the fundamental procedure given below:

Step 1. Confirm the contact tip

- Confirm the inner-diameter indication on the contact tip matches the diameter of the welding wire to be used and the bore of the contact tip is not deformed by abrasion.
- Confirm the contact tip is firmly fastened in place so that welding currents can surely be conveyed from the contact tip to the welding wire during welding.

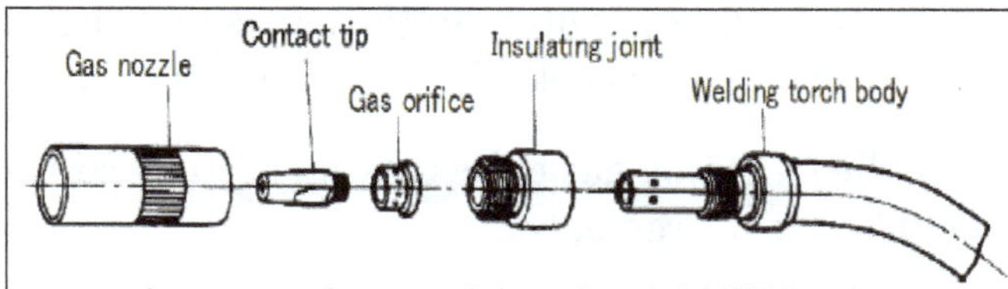

Arrangement of a contact tip in an air-cooled GSMAW torch

Step 2. Confirm the feed roller

- Confirm the groove size of the feed roller matches the diameter of the welding wire to be used.
- If the groove of the feed roller is worn out, replace it with a new one.

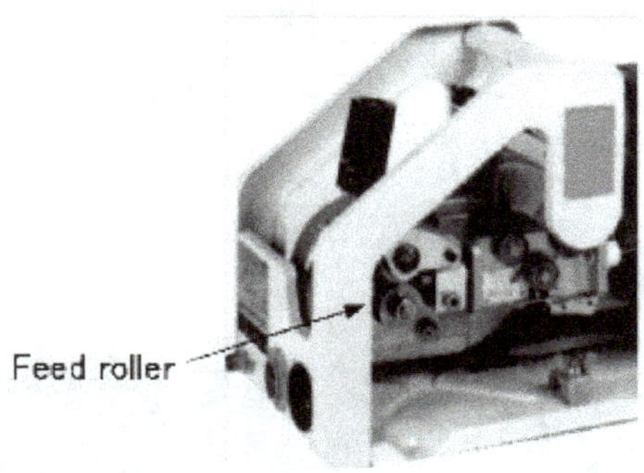

Arrangement of a feed roller in a wire feeder

Step 3. Turn on the power switches

- Confirm the main switch of the power source is turned off, and turn on the main switch of the power line switchboard.
- Confirm the torch switch is turned off, and turn on the main switch of the power source, then confirm the power pilot lump turns on.
- Confirm the blower of the power source rotates smoothly.

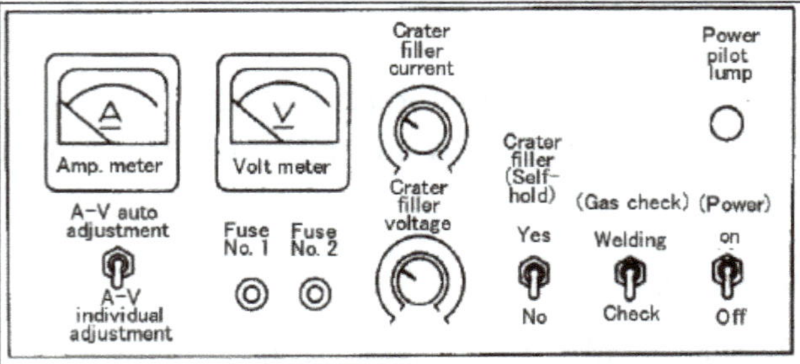

A typical control panel of a GSMAW power source

Step 4. Adjust the flow rate of a shielding gas

- Confirm the shielding gas regulator is fitted in the correct position and the build-in heater is working, before opening the main cock of a shielding gas cylinder.
- Adjust the pressure of the shielding gas to be 0.2-0.3 MPa by controlling the pressure regulation knob of the shielding gas regulator.
- Turn the gas check switch to the "Check" position on the control panel of the power source and adjust the flow rate of the shielding gas by controlling the knob on the gas flow meter according to the standard rates shown below.
- Return the gas check switch to the "Welding" position.

Welding Current (A)	Gas flow rate (liter/min)
100-200	15-25
200-300	20-30
300-500	20-30

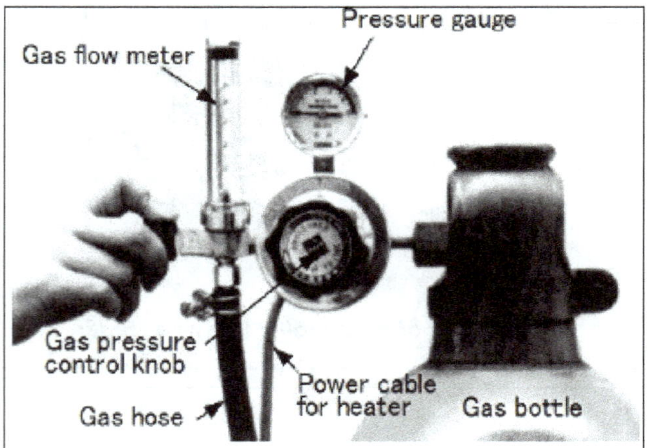

A gas regulator

Step 5. Set a welding wire on the spindle

- Set a specified welding wire onto the spindle of the wire feeder so that the tip of the spooled wire can be taken out under the spool toward the feed roller and set the spool stopper.

- When handling a spooled wire, take the correct way as shown below. The wrong way causes deformation of the flange of the spool and may cause the wire insertion between layers of wires, which finally may cause the wire jamming during welding.

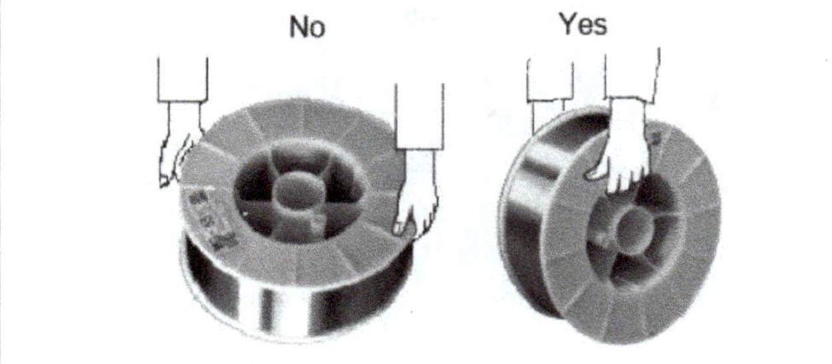

A correct way (Right) and a wrong way (Left) of lifting a spool of wire

Step 6. Engage the wire with the feed roller

- Take off the tip of the spooled wire from the stopper hole, cut off the tip, holding the wire by hand, and straighten a certain length of the wire by hand for easier guidance to the wire inlet guide. Lead the wire into the wire inlet guide through the feed roller, and confirm the wire aligns with the groove of the feed roller. Set the pressure arm down onto the wire to engage the wire with the feed roller.
- Adjust the pressure on the wire by controlling the pressure-adjusting knob according to the size and material of the wire to be used. Excessive Pressure causes deformation of the wire and flakes of the wire between the feed roller and the pressure roller, thereby causing irregular wire feeding. In contrast, too weak pressure causes slipping of the wire, thereby causing unstable wire feeding.

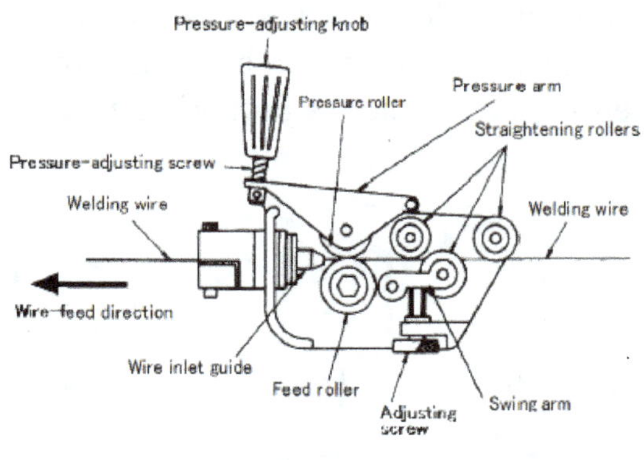

Alignment of the straightening roller, feed roller, wire inlet guide, and a welding wire (Note: Wire straightening rollers are needed particularly for automatic and robotic welding)

Step 7. Feed the wire into the welding torch

- By turning on the inching switch of the remote controller, feed the wire through the welding torch until the wire extends approximately 20mm from the tip of the contact tube.
- Confirm the wire is fed smoothly. If the wire feeding is not smooth, check whether or not

the pressure roller properly presses the wire, the feed roller and the wire are aligned and the wire spool is smoothly rotated.

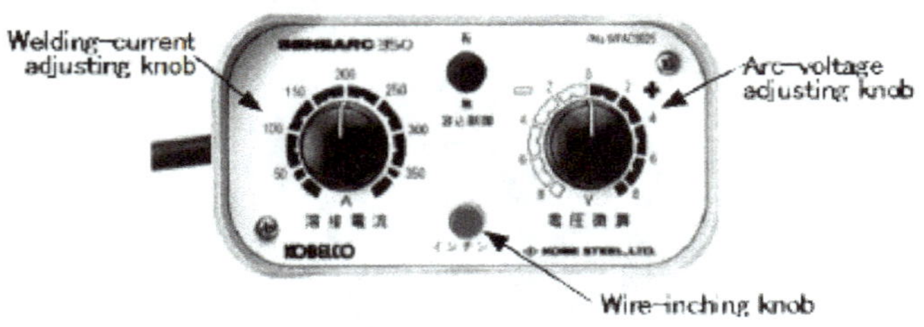

A remote controller for a GSMAW power source

Step 8. Adjust the straightness of the wire

- Adjust straightness of the wire by controlling the adjusting screw of the swing arm to feed the wire properly straight from the contact tip and to decrease the feeding resistance of the wire passage.

Step 9. Adjust welding current and voltage

- Adjust the welding current and voltage by controlling the welding-current adjusting knob and the arc-voltage adjusting knob of the control panel of the power source or of the remote controller. Proper welding current and arc voltage depend on the size of the wire, thickness of base metal, and welding position.
- With the proper source having the A-V automatic control system, the arc voltage is automatically adjusted when the welding current is set, according to the pre-set A-V relationship. More fine adjustment can be done by using the voltage fine – adjusting knob.

Note: If the voltage is set too low, the welding wire sticks onto the base metal making crackling sounds with an irregular arc. If the voltage is set too high, the arc becomes a long flame causing much spatter generation. When the arc voltage is properly adjusted, the arc is stable with little spatter generation.

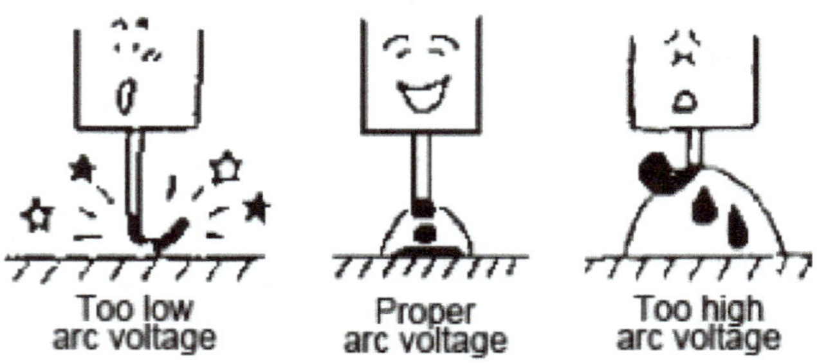

Adjustment of proper arc voltage

Step 10. Holding welding torch in the proper position and start welding

- Hold, the welding torch in the proper position suitable for controlling the molten pool, maintaining a stable arc and good gas – shielding.

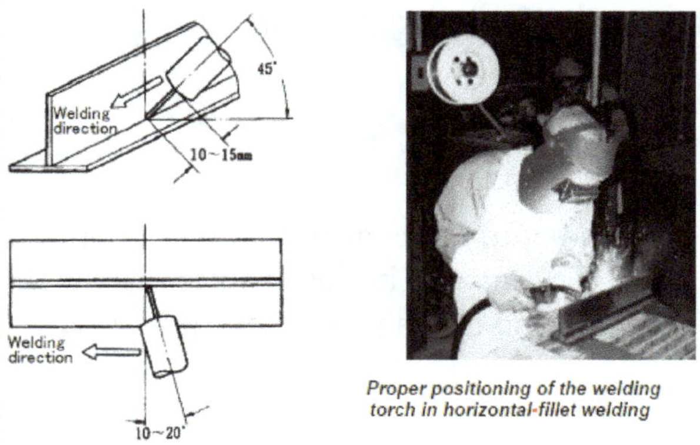

Proper positioning of the welding torch in horizontal-fillet welding

- Confirm the conduit cable has no sharp bend of small radius because a sharp bend of the conduit cable causes irregular wire feeding, thereby causing an irregular arc. The permissible minimum bend is approximately 150 mm radius for a curvature or 300 mm diameter for a circle as shown in below figure.

Permissible minimum radius of a curvature and diameter of a circle in a conduit cable line to prevent irregular wire feeding and an unstable arc

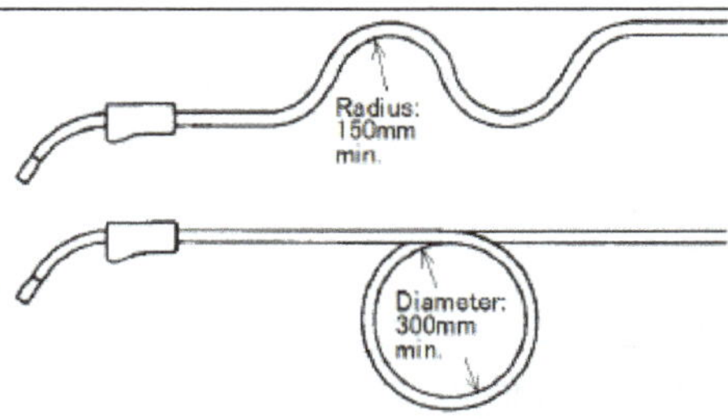

- Hold the welding torch so that the tip of the wire is close to the base metal, turn on the torch switch to feed the shielding gas and welding wire, then an arc starts when the wire touches the workpiece. While the welder holds the torch switch (turned on) the arcing continues, and when the welder releases the torch switch (turned off) the arc turns off, with a regular electrical control circuit.

In contrast, with the torch trigger circuit for the "Self Holding Function" once the torch switch is pulled in (turned on), the acing continues even if the welder releases the torch switch. This function is useful to release the welder from fatigue of holding the torch switch during welding for a long welding line. For termination of the weld bead, the welder has to pull the torch switch in for crater treatment, "Crater Filler Function," followed by releasing the torch switch to cut the arc. The crater filling current and voltage can be initially set on the control panel of the power source.

Step 11. Close the welding operation

- Close the main cock of the shield gas cylinder, discharge the shielding gas remained in the gas passage between the gas regulator and the welding torch by operating the gas check switch. After confirming the gas regulator indicates the zero pressure, return the gas switch to "Welding".
- Turn off the main switch of the power source.
- Turn off the main switch of the power line switchboard.

CHAPTER 2

Equipments used in Shielded Metal Arc Welding

Shielded metal arc welding process requires a constant power supply, an electrode and electrode holder, and welding cables. The topics covered in this chapter discuss the various types of equipment and the techniques of checking and choosing the best quality equipment for use in shielded metal arc welding.

Power Sources

Shielded Metal Arc Welding (SMAW) is the terminology approved by the American Welding Society.

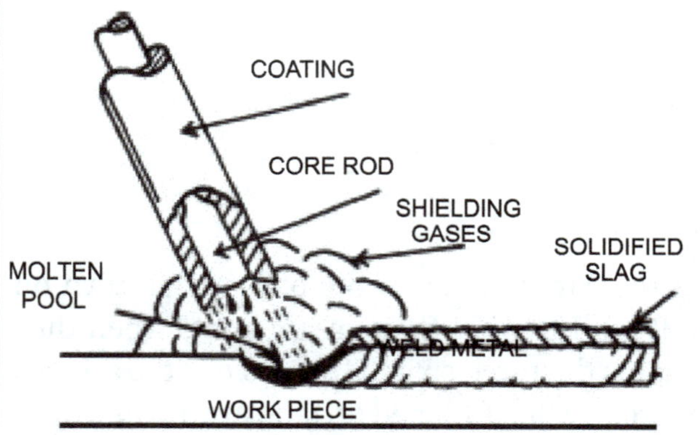

SHIELDED METAL ARC WELDING

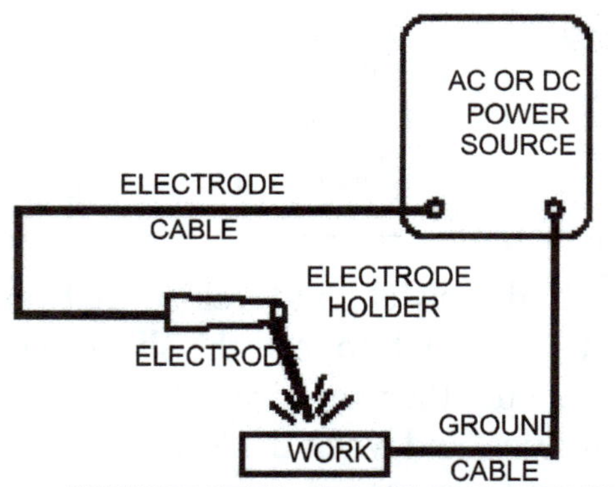

SHIELDED METAL ARC WELDING CIRCUIT

Equipment and Operation

One reason for the wide acceptance of the SMAW process is the simplicity of the necessary equipment. The equipment consists of the following items:

1. Welding power source
2. Electrode holder
3. Ground clamp
4. Welding cables and connectors
5. Accessory equipment (chipping hammer, wire brush)
6. Protective equipment (helmet, gloves, etc.)

Welding Power Sources

Shielded metal arc welding may utilize either alternating current (AC) or direct current (DC), but in either case, the power source selected must be of the constant current type. This type of power source will deliver a relatively constant amperage or welding current regardless of arc length variations by the operator. The amperage determines the amount of heat at the arc and since it will remain relatively constant, the weld beads produced will be uniform in size and shape.

The primary function of a welding power source is to provide sufficient power to melt the joint. However with MMA the power source must also provide current for melting the end of the electrode to produce weld metal, and it must have a sufficiently high voltage to maintain the arc. A constant current (drooping) characteristic is used.

MMA electrodes are designed to be operated with alternating current (AC) and direct current (DC) power sources. Although AC electrodes can be used on DC, not all DC electrodes can be used with AC power sources.

As MMA requires a high current (50-300A) but a relatively low voltage (10-50V), high voltage mains supply (240 or 440V) must be reduced by a transformer. To produce DC, the output from the transformer must be further rectified. To reduce the hazard of electrical shock, the power source must function with a maximum no-load voltage, that is, when the external (output) circuit is open (power leads connected and live) but no arc is present. The no-load voltage rating of the power source is as defined in IEC 60974-1 and IEC 60974-6 and must be in accordance with the type of welding environment or hazard of electrical shock. The power source may have an internal or external hazard reducing device to reduce the no-load voltage; the main welding current is delivered as soon as the electrode touches the workpiece. For welding in confined spaces, you should use DC or AC with a low voltage safety device to limit the voltage available at the holder to less than 48V rms.

There are four basic types of power source:

- AC transformer
- DC rectifier
- AC/DC transformer-rectifier
- DC generator

AC electrodes are frequently operated with the simple, single phase transformer with current adjusted by means of tappings or sliding core control. DC rectifiers and AC/DC transformer-rectifiers are controlled electronically, for example by thyristors. Modern power sources called inverters use transistors to convert mains AC (50Hz) to a high frequency AC (typically 50 kHz) before transforming down to a voltage suitable for welding and then rectifying to DC. Because high frequency transformers can be relatively small, principal advantages of inverter power sources are undoubtedly their size and weight when the source must be portable.

Power Supply Designs

The welding power supplies most commonly seen can be categorized within the following types.

Transformer

A transformer-style welding power supply converts the moderate voltage and moderate current electricity from the utility mains (typically 230 or 115 VAC) into a high current and low voltage supply, typically between 17 and 45 (open-circuit) volts and 55 to 590 amperes. A rectifier converts the AC into DC on more expensive machines.

This design typically allows the welder to select the output current by variously moving a primary winding closer or farther from a secondary winding, moving a magnetic shunt in and out of the core of the transformer, using a series saturating reactor with a variable saturating technique in series with the secondary current output, or by simply permitting the welder to select the output voltage from a set of taps on the transformer's secondary winding. These transformer style machines are typically the least expensive.

The trade off for the reduced expense is that pure transformer designs are often bulky and massive because they operate at the utility mains frequency of 50 or 60 Hz. Such low frequency transformers must have a high magnetizing inductance to avoid wasteful shunt currents. The transformer may also have significant leakage inductance for short circuit protection in the event of a welding rod becoming stuck to the workpiece. The leakage inductance may be variable so the operator can set the output current.

Generator and Alternator

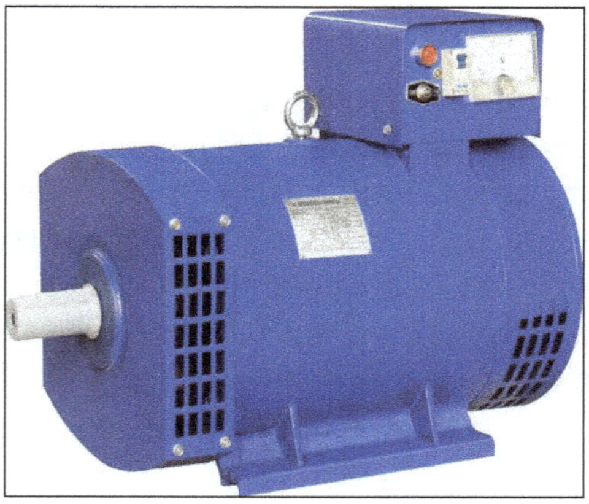

Welding power supplies may also use generators or alternators to convert mechanical energy into electrical energy. Modern designs are usually driven by an internal combustion engine but older machines may use an electric motor to drive an alternator or generator. In this configuration the utility power is converted first into mechanical energy then back into electrical energy to achieve the step-down effect similar to a transformer. Because the output of the generator can be direct current, or even a higher frequency AC, these older machines can produce DC from AC without any need for rectifiers of any type, or can also be used for implementing formerly-used variations on so-called heliarc (most often now called TIG) welders, where the need for a higher frequency add-on module box is avoided by the alternator simply producing higher frequency ac current directly.

Inverter

Since the advent of high-power semiconductors such as the insulated gate bipolar transistor (IGBT), it is now possible to build a switched-mode power supply capable of coping with the high loads of arc welding. These designs are known as inverter welding units. They generally first rectify the utility AC power to DC; then they switch (invert) the DC power into a stepdown transformer to produce the desired welding voltage or current. The switching frequency is typically 10 kHz or higher. Although the high switching frequency requires sophisticated components and circuits, it drastically reduces the bulk of the step down transformer, as the mass of magnetic components (transformers and inductors) that is required for achieving a given power level goes down rapidly as the operating (switching) frequency is increased. The inverter circuitry can also provide features such as power control and overload protection. The high frequency inverter-based welding machines are typically more efficient and provide better control of variable functional parameters than non-inverter welding machines.

The IGBTs in an inverter based machine are controlled by a microcontroller, so the electrical characteristics of the welding power can be changed by software in real time, even on a cycle by cycle basis, rather than making changes slowly over hundreds if not thousands of cycles. Typically, the controller software will implement features such as pulsing the welding current, providing variable ratios and current densities through a welding cycle, enabling swept or stepped variable frequencies, and providing timing as needed for implementing automatic spot-welding; all of these features would be prohibitively expensive to design into a transformer-based machine, but require only program memory space in a software-controlled inverter machine. Similarly, it is possible to add new features to a software-controlled inverter machine if needed, through a software update, rather than through having to buy a more modern welder.

Other Types

Additional types of welders also exist, besides the types using transformers, motor/generator, and inverters. For example, laser welders also exist, and they require an entirely different type of welding power supply design that does not fall into any of the types of welding power supplies discussed previously. Likewise, spot welders require a different type of welding power supply, typically containing elaborate timing circuits and large capacitor banks that are not commonly found with any other types of welding power supplies.

Controls

The controls are usually located on the front panel of the welding machine. These usually consist of a knob or tap switch to set the rough current range and a knob to adjust the current within the set range. On DC welding machines there is usually a switch to change polarity, and on combination AC-DC machines, there is usually a switch to select the polarity or AC current. An On-Off switch is also located on the front of the machine.

Arc Force Control is a function of amperage triggered by a preset (internal module) voltage. The preset trigger voltage is 18 volts. What this means is that anytime the arc voltage drops from normal welding voltage to 18 volts or less, the drop triggers the arc force current, which gives the arc a surge of current to keep the arc from going out.

When an arc is struck, the electrode is scratched against the work. At that point, the voltage goes to -0- which triggers the arc force current and the arc is initiated quickly. On a standard machine without arc force control, arc striking is difficult and electrode sticking may occur.

After the arc is established, a steady burn-off is desired. As the electrode burns and droplets of metal are transferred from the end of the electrode to the work piece, there is a time period when the droplet is still connected to the end of the electrode but is also touching the work piece. When this occurs, the machine is, in effect, in a "dead-short" - the voltage drops, the arc force is triggered and the droplet is transferred. On machines without arc force, the burn-off is the same; however, without the arc force to help, an arc outage may occur, and the electrode will stick in the puddle.

In tight joints, such as pipe welding, the arc length is very short and with standard machines, it is difficult to maintain the arc since it wants to "short-out" against the sidewalls or bottom of the joint. The arc force control can be adjusted on this type application to prevent electrode sticking; whenever the voltage drops, the drop triggers the arc force current and the sticking doesn't happen because the current surge occurs.

In many applications, there is a need for a very forceful arc to obtain deeper penetration, or in the case of arc gouging, the forceful arc is essential in helping to force the metal out of the groove being gouged. With arc force control, this type application is made much easier than with conventional machines, with which arc length becomes critical and arc outages can occur.

When welding with a given size electrode, there is always an optimum amperage setting. When using arc force control, the optimum amperage setting is continually working to maintain the arc, which means that although we can't see it on the meters, there is usually some added amperage to assist in rod burn-off. This in turn means we really get a slightly faster burn-off than with a conventional rectifier.

When working out-of-position, a forceful arc is needed to help put metal in place. Each individual operator can adjust the arc force control to provide just the amount needed. Arc force can also be of assistance when welding rusty or scaly material, since the more forceful arc will help to break up these deposits.

Electrode Holder

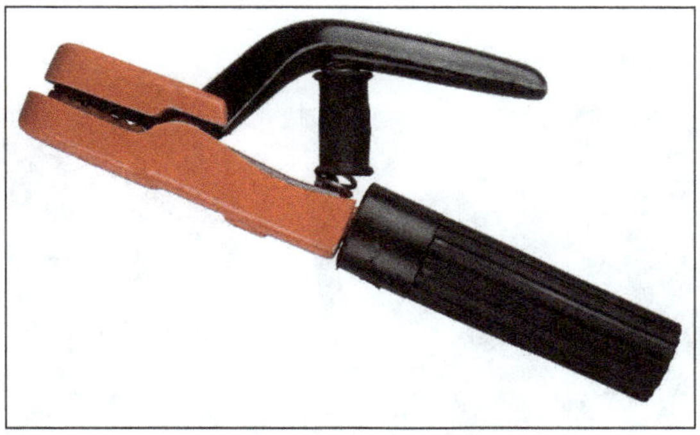

An electrode holder, commonly called a stinger, is a clamping device for holding the electrode securely in any position. The welding cable attaches to the holder through the hollow insulated handle. The design of the electrode holder permits quick and easy electrode exchange. Two general types of electrode holders are in use: insulated and noninsulated. The noninsulated holders are not recommended because they are subject to accidental short circuiting if bumped against the work- piece during welding. For safety reasons, try to ensure the use of only insulated stingers on the jobsite.

Electrode holders are made in different sizes, and manufacturers have their own system of designation. Each holder is designed for use within a specified range of electrode diameters and welding current. You require a larger holder when welding with a machine having a 300-ampere rating than when welding with a 100-ampere machine. If the holder is too small, it will overheat.

How to make an Electrode Holder

Step 1. Material

For this, you will need

- 3/8 brass plumbing end plug (size can vary if you are making a larger device)
- 13/32 (or 7/16), 7/32, and 1/8 bits

- all my measurements assume this is going on a 1 AWG cord. By all means, adjust for your cord and electrode size
- 8-32 knurled brass bolt
- 8-32 tap
- AWG 10 single core copper wire
- soldering torch
- solder with rosin core, try to go lead free
- some shrink tube. You will need a few sizes to cover all the parts,s o make sure you have large enough ones.

Step 2. Drill the Back Hole

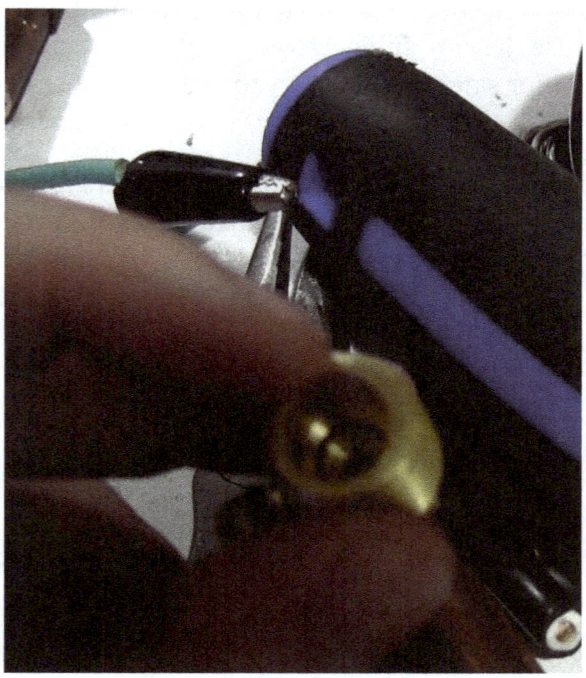

The first hole you want to make is the 13/32th hole in the back.

Step 3. Solder the Wire in

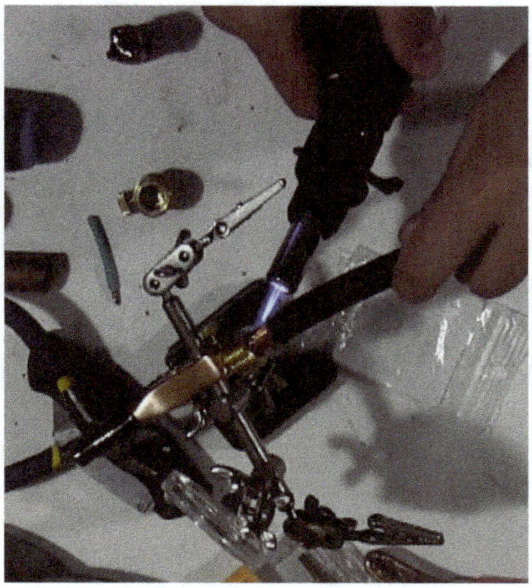

If you make the other holes now, you will have a to redo part of it as the solder will run into the other holes. If you make your holes shallow enough, you can drill them all before, as long as the back does not connect with the pinching screw's hole or the electrode's hole. Either you get ease of clamping, or ease of depth gauging.

Step 4. Drill some more

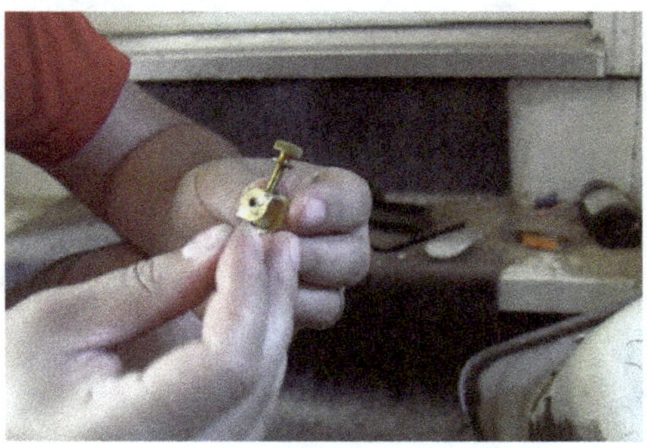

Drill the side 1/8th hole for the 8-32 finger screw that will hold the electrode, and the the 7/32 hole from the tip inwards.

Step 5. Tap your 1/8th Hole

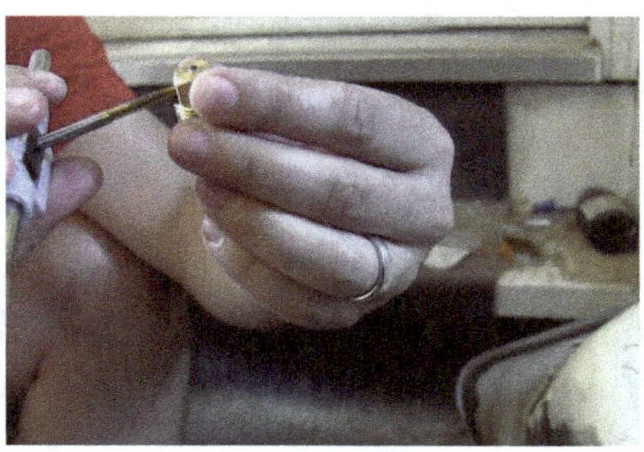

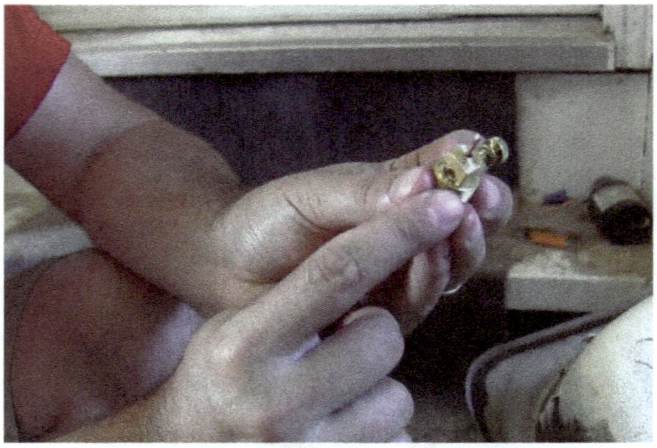

Tap the hole for the finger screw to actually screw in, and that's it, you are done. Everything is ready for your use.

Step 6. Finishing Touch

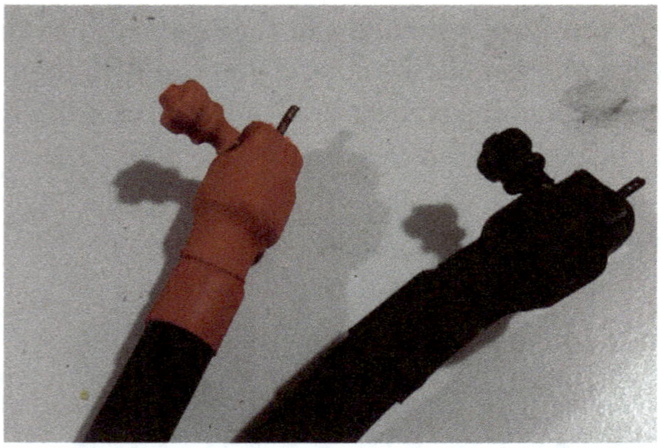

You can also add some shrink tube to the whole thing. It serves to prevent accidental contact, but also colour codes your cables so you know what is plugged where.

Shielded Metal Arc Welding Electrode

Shielded arc or heavy coated welding electrodes have a definite composition on which a coating has been applied by dipping or extrusion. The electrodes are manufactured in three general types: those with cellulose coatings; those with mineral coatings; and those whose coatings are combinations of mineral and cellulose. The cellulose coatings are composed of soluble cotton or other forms of cellulose with small amounts of potassium, sodium, or titanium, and in some cases added minerals. The mineral coatings consist of sodium silicate,metallic oxides clay, and other inorganic substances or combinations thereof. Cellulose coated electrodes protect the molten metal with a gaseous zone around the arc as well as the weld zone. The mineral coated electrode forms a slag deposit. The shielded arc or heavy coated electrodes are used for welding steels, cast iron, and hard surfacing.

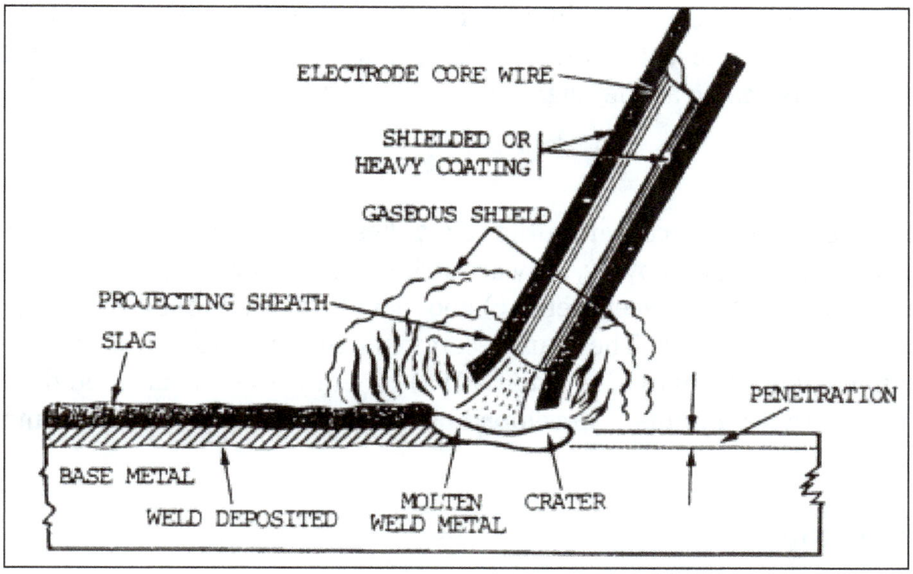

Arc Action obtained with a Shielded Arc Electrode

Functions of Shielded Arc or Heavy Coated Electrodes

These welding electrodes produce a reducing gas shield around the arc. This prevents atmospheric oxygen or nitrogen from contaminating the weldmetal. The oxygen readily combines with the molten metal, removing alloying elements and causing porosity. Nitrogen causes brittleness, low ductility, and in Some cases low strength and poor resistance to corrosion.

They reduce impurities such as oxides, sulfur, and phosphorus so that these impurities will not impair the weld deposit.

They provide substances to the arc which increase its stability. This eliminates wide fluctuations in the voltage so that the arc can be maintained without excessive spattering.

By reducing the attractive force between the molten metal and the end of the electrodes, or by reducing the surface tension of the molten metal, the vaporized and melted coating causes the molten metal at the end of the electrode to break up into fine, small particles.

The coatings contain silicates which will form a slag over the molten weld and base metal. Since the slag solidifies at a relatively slow rate, it holds the heat and allows the underlying metal to cool and solidify slowly. This slow solidification of the metal eliminates the entrapment of gases within the weld and permits solid impurities to float to the surface. Slow cooling also has an annealing effect on the weld deposit.

The physical characteristics of the weld deposit are modified by incorporating alloying materials in the electrode coating. The fluxing action of the slag will also produce weld metal of better quality and permit welding at higher speeds.

The steel electrode of a given composition is made available with different types of flux coating in order to make them suitable for different arc characteristics, welding position, welding speed, deposition rate, weld metal recovery, weld metal properties and variety of quality requirements. The selection of correct type of electrode coating results in weld metal with desired quality characteristics at low cost. In general, welding electrode is selected in such a way that characteristics of weld metal are similar to or better than the base material while keeping in mind the welding position and weld joints design as they significantly affect the properties of the weld.

Rutile Electrode

These electrodes predominantly contain rutile (TiO_2) besides other constituents and are known to offer almost 100% weld metal recovery, easy arc striking and restriking. These are found suitable for a) fillet welds, b) welding of sheet metal, c) good gap bridging capability, d) free from spatter losses and e) all position welding. These are recommended for welding low strength steel (<440 MPa). For welding of high strength steel (>440 MPa) generally weld metal should have low hydrogen level and therefore weld joints is developed using basic, rutile, basic-rutile and Zirconbased electrode.

Cellulosic Electrodes

These electrodes are composed of large amount of hydrocarbon compounds and calcium carbonates besides other constituents and are found suitable for a) all welding positions especially for vertical and overhead welding position and b) realizing high mechanical properties in a weld metal of radiographic quality. These are preferred for vertical downward welding. However, these produce high hydrogen content in weld metal besides deep penetration.

Acidic Electrode

Acidic electrodes offer a) easier arc striking than basic electrodes but poorer arc striking than rutile electrodes, b) moderate welding speed, c) smooth weld bead d good slag detachability. However, acidic electrode has been replaced by rutile electrode and basic electrode for flat and positional welding respectively. The ductility and toughness weld metal developed by acidic electrode are better than those developed from rutile electrodes however yield and ultimate tensile strength are found inferior. This type of electrode results in minimal penetration which is good for very thin sheet but these are sensitive to moisture pick up.

Basic Electrode

These electrodes have basic (alkali) coatings containing calcium carbonate / calcium fluoride.The basic electrodes are preferred over other electrode for developing weld joints of high strength steel (480-550 MPa) with weld metal having a) low hydrogen, b) good low temperature toughness, c) resistance to hot and cold cracking. However, these electrodes suffer from comparatively poor slag detachability. The welding speed and deposition rate offered by the basic electrodes especially in vertical welding position is much higher than the rutile and acidic electrode. Basic electrodes can sustain higher welding current even in vertical welding position.

Basic-rutile Electrode

This type of electrode combines positives of both basic as well as rutile electrode and therefore recommended for horizontal–vertical fillet welds of high strength steels.

How to Select the Right Stick Electrode

Stick welding is a skill that takes much practice to master, and selecting the correct electrode for the job can be equally challenging. Stick electrodes are available in a wide range of types, each of which provides different mechanical properties and operates with a specific type of welding power source. Factors you should consider when selecting a stick electrode include base metal type, joint fit-up and welding positions. Before you power up your machine and pick up your electrode holder, consider these basic guidelines.

Assess your Base Metal

The first step in choosing an electrode is to determine your base metal composition. Your goal is to match (or closely match) the electrode composition to the base metal type, which will help ensure a strong weld. If you're in doubt about the composition of your base metal, ask yourself these questions:

- What does the metal look like?

If you're working with a broken part or component, check for a coarse and grainy internal surface, which usually means the base material is a cast metal.

- Is the metal magnetic?

If the base metal is magnetic, chances are good that the base metal is carbon steel or alloy steel. If the base metal is not magnetic, the material could be manganese steel, 300 series austenitic stainless steel or a non-ferrous alloy such as aluminum, brass, copper or titanium.

- What kind of sparks does the metal give off when touched by a grinder?

As a rule of thumb, more flare in the sparks indicates a higher carbon content such as in A-36 grade steel.

- Does a chisel "bite" into the base metal or bounce off?

A chisel will bite into a softer metal, such as mild steel or aluminum, and bounce off of harder metals, such as high carbon steel, chrome-moly or cast iron.

Tensile Strength

To prevent cracking or other weld discontinuities, match the minimum tensile strength of the electrode to the tensile strength of the base metal. You can identify a stick electrode's tensile strength by referring to the first two digits of the AWS classification printed on the side of the electrode. For example, the number "60" on an E6011 electrode indicates that the filler metal produces a weld bead with a minimum tensile strength of 60,000 psi and, as a result, would work well with a steel of similar tensile strength.

Welding Current

Some electrodes can be used with only AC or DC power sources while other electrodes are compatible with both. To determine the correct current type for a particular electrode, refer to the fourth digit of the AWS classification, which represents the type of coating and type of compatible welding current.

Fourth Digit	Type of Coating	Welding Current
0	Cellulose Sodium	DCEP
1	Cellulose Potassium	AC or DCEP
2	Titania Sodium	AC or DCEN
3	Titania Potassium	AC, DCEP or DCEN
4	Iron Powder Titania	AC, DCEP or DCEN
8	Iron Powder Low Hydrogen	AC or DCEP

DCEP – Direct Current Electrode Positive DCEN – Direct Current Electrode Negative

Refer to the fourth digit of the AWS classification to determine the compatible welding current

The type of current you use also influences the penetration profile of the resulting weld. For example, a DCEP compatible electrode, such as an E6010 delivers deep penetration and produces an extremely tight arc. It also has the ability to "dig" through rust, oil, paint and dirt. A DCEN compatible electrode, such as an E6012, provides mild penetration and works well when bridging two joints or welding high speed, high current fillet welds in the horizontal position.

An AC compatible electrode, such as an E6013, produces a soft arc with medium penetration and should be used to weld clean, new sheet metal.

Base Metal Thickness, Shape and Joint Fit-Up

Thick materials require an electrode with maximum ductility and low hydrogen to prevent weld cracking. Electrodes with AWS classification numbers ending in 15, 16 or 18 provide excellent low–hydrogen properties and good toughness (high impact values) to accommodate for residual stress.

For thin materials, you will need an electrode that produces soft arcs, such as a 6013, Also, smaller diameter electrodes will provide shallow penetration to help prevent burn-through on thinner materials.

You'll also want to assess the joint design and fit-up. If you're working on a joint with a tight fit-up or one that is not beveled, use an electrode that provides a digging arc to ensure sufficient penetration, such as an E6010 or E6011. For materials with wide root openings, select an electrode, such as an E6012, that creates a concave weld face suitable for bridging gaps and making groove welds.

Welding Position

To determine what position(s) a particular electrode is qualified for, refer to the third digit in AWS classification. Here's how you decipher the qualified electrode position:

1 = flat, horizontal, vertical and overhead

2 = flat and horizontal only

For example, a 7018 electrode can be used in the flat, horizontal, vertical and overhead positions.

Specification and Service Conditions

Make sure to assess the conditions that the welded part will encounter throughout its service. If it will be used in high heat or low temperature environments, subjected to repetitive shock loading, a low hydrogen electrode with higher ductility will reduce the chance of weld cracking. Also, be certain to check for welding specifications if you're working on critical applications such as pressure vessel or boiler fabrication. In most cases, these welding specifications will require you to use specific types of electrodes.

Environmental Job Conditions

To achieve the best results, you should always remove excessive mill scale, rust, moisture, paint and grease. Clean base metals help prevent porosity and increase travel speeds. If cleaning your base metal is not possible, E6010 or E6011 electrodes deliver a deep penetrating arc that has the ability to cut through contaminants.

How to Check the Quality of SMAW Electrodes before Fabrication

Shield metal arc welding (SMAW) electrodes are the most popular filler metals in the world, offering the widest range of alloy types and brand available in the global market. Each country has its own SMAW electrode specifications; however, American Welding Society (AWS) specifications are widely accepted. Carbon steel SMAW electrodes, for example, have two major AWS specifications,

AWS A5.1, for mild carbon steel electrodes, and AWS A5.5 for low-alloy steel electrodes. These specifications classify electrodes per the following:

- Tensile strength grade of a deposit metal
- Micro-alloying elements concentrations of a deposit metal
- Diffusible hydrogen content of a deposit metal
- Welding electrical characteristics, i.e. AC/DC
- Flux coating type
- Positional welding capability

The classification format and designators are detailed in every AWS A5 specification. Figure below illustrates the classification of a 70ksi mild carbo steel electrode.

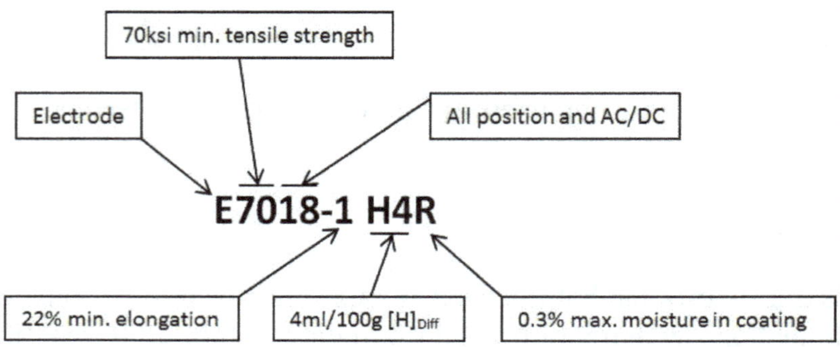

AWS classification of a 70ksi mild carbon steel electrode

While SMAW electrodes manufacturing specifications are specifically designed by filler metal companies to reliably produce quality electrodes that satisfy the requirements of AWS A5 specifications, it is always important to confirm electrode quality in the field to ensure stream-lined fabrication. The required steps are as follows:

1. Packaging Examination

SMAW electrodes are usually packaged in cans or tubes, and some critical types of electrodesare also vacuum-sealed in a hermetic foil. After a batch of SMAW electrodes is received, a sample pack should be randomly picked from the shipment. The packaging integrity should first be examined to identify any damage, as a broken package could result in moisture absorption by the electrodes. For example, the flux coating of a "H4R"-designated electrode can absorb moisture from the environment, causing the moisture content to exceed the specified 0.3% maximum upper limit. Consequently, this package may not be able to satisfy the H4 requirement of low-hydrogen electrodes. This excessive moisture pick-up may also generate a large amount of spatter, and the resulting spatter clean-up will add time and cost to the fabrication operation. Moisture pick-up may also negatively affect arc stability, and may result in porosity. In extreme cases, rusty stubs of electrodes may be found, which should prohibit the use of the entire batch.

2. Surface Quality Examination

Most SMAW electrodes are manufactured by an extrusion process. While it is rare to manufacture electrodes using the classic dipping process, this method is still used by a few manufacturers to produce special electrodes. The extrusion production consists of five major steps:

- Dry flux mixing
- Wet mix blending
- Cylindrical dough pressing
- Electrode extrusion
- Electrodes drying and baking

Usually, potassium and/or sodium silicate liquid(s) are blended with a dry flux mix. It is critical to use a right amount of silicate liquid(s) with a controlled viscosity to achieve uniform wet mix blending. The wet mix is then pressed to generate cylindrical dough to be loaded in an extruding chamber. The extrudability of the dough is critical, and is directly determined by the wet mix blending process. The flux coating is then extruded onto the core wire under a pressure between 3000-5000 psi.

The quality of an extrusion process is evident on the surface of the electrode flux coating. A twisted flux coating is an indication that the electrode may have been too wet. This may be caused by an excessive amount of silicate liquid(s) blended in the wet mix. This twist may result in eccentricity between the core wire and the flux coating, which may then negatively affect the stability of the welding arc. As such, a handful of electrodes must always be visually examined to check for this issue.

When checking for eccentricity at the tip of an electrode, attention should also be paid to the integrity of the flux coating. If the flux coating is broken at the tip, it may indicate that the flux coating or the bonding between the core wire and coating is not strong enough. The strength of the coating is directly related to the flux formulation and the binder(s) used to make the wet mix. Some filler metal manufacturers have proprietary coating strength testers for QA control. If cracks are found on an electrode coating, it indicates a poor electrode baking process. Generally speaking, flux coating defects are indicative of quality issues.

Flux formulas are often adjusted to reduce cost. A change in the coating color may result, indicating an alteration or re-formulation of the raw materials used. If abnormal welding behavior is observed when using a colorchanged electrodes, it should be reported without delay.

3. Labels

Usually, the AWS classification of an electrode is imprinted on the coating near the electrode stub. It is important to check the imprinted label on the electrode and the product label shown on package with respect to the purchase order to prevent the wrong electrode from being used.

It is a typical practice to use H4 low hydrogen electrodes to make critical seam-welds. Therefore, it is important to check the diffusible hydrogen designator imprinted on an electrode coating and a product label to prevent welding with an electrode with the wrong diffusible hydrogen designation.

Similarly, it is a good practice to check the labels to distinguish a low-carbon graded stainless-steel electrode (with a "L" designator behind the composition designators) from a regular-carbon graded stainless-steel electrode.

4. Straightness of the Electrode

Before electrode extrusion, the core wires are sized down by a die train from a large-diameter bare-wire coil. They are then cut in length after a wire straightening operation. Occasionally, some core wires are not straightened well and may be slightly bent. When those bent wires pass through an extrusion die, eccentricity may result. The electrode could also be bent by a non-uniform pressure distribution on the dough in the extrusion chamber. This is more prevalent in extruders which use vertical pressing. Electrodes can be effectively examined by rolling them on a table top, as excessive wobbling indicates a lack of straightness. The eccentricity which results from bent electrodes may negatively affect the welding performance.

Welding Cables

Welding cable is a popular portable cord that is used in various welding applications as well as many power supply applications not exceeding 600V where some flexibility is required.

Welding cable sizes range from 6 awg to 500 MCM. It consists of bare annealed copper per ASTM B-3. The jacket on most welding cable is thermoset, typically EPDM or Neoprene. The most widely manufactured colors of standard welding cable are black and red. The standard temperature rating 90C.

Many times welding cables are used in applications outside of welding but it is dependent on your specific requirements. It is mainly designed for use with connections from electrode holder and clamp to arc welder, welding box, bus or transformer. We always recommend checking the welding cable ampacities chart prior to using welding cable for a welding application or a 600V in-line application.

There are many manufactures of welding cable but the two of the largest are General Cable and American Insulated Wire. Both of these manufactures also manufacture more durable versions of the standard welding cable for extra hard usage. These will typically include a brighter colored jacket (usually orange), have higher strand counts for increased flexibility, higher temperature rating and greater industry approvals. Be aware these higher grade welding cables are rarely available on put-up lengths over 500′. Many times these more durable versions of welding cable can be used in applications such as on a hoist or crane to save money over the more expensive pendant and reeling cable. You should always check with your engineer for exact requirements of your electrical job before you use a welding cable in an application outside of its normal specifications.

It is important to understand the difference in strand counts on your welding cable. The stranding is what makes up the majority of the flexibility of the cable. Many companies market their standard welding cable as Super Flex or Ultra Flex but they are selling a normal strand count welding cable.

There are two primary types of stranding on a welding cable. One is Class K, which is 30 awg strands and this will be your typical welding cable that is offered most predominately in black and red jacket colors. The other version is Class M stranding, and that is 34 awg and is a more durable and flexible version. It is the one that typically comes in the brighter florescent jacket and it will cost 20-40% more than typical welding cable due to the increase in copper involved and tougher jacketing properties. You should always ask the type of stranding on the welding cable you are being sold so you are aware of the exact product you are ordering.

How to Choose the Right Welding Cables

1. Ampacity

Determining the strength of the electrical current you'll be using on a job site is a very important first step. Once you've got a range for the currents you will need, you can begin to shop for a welding cable based on its ampacity rating. Ampacity stands for "ampere capacity," and the rating of various cables will narrow down your list of options considerably. This is probably the most vital of all of the factors to consider, since you absolutely cannot safely run higher currents through wires that aren't built to handle it.

2. Insulation

The grade of insulation of a cable is also a determining factor on which cable you choose. If your work environment is low-hazard, you might be fine with a more thinly insulated cable. However, if there is significant risk of moisture, chemicals, oils, or heat in your work environment, you will need to choose a cable with a strong insulation layer that is rated to be resistant to the hazards of your workplace. This is especially important when welding in marine and garage environments, where oil and moisture are commonplace. Both salt water and oil can erode standard grade insulation and lead to dangerous workplace conditions and tool failure.

3. Flexibility

You will no doubt need flexible cables if you are going to be working in tight spaces or at challenging angles. The downside of cable flexibility is that it is affected by both the insulation and the thickness of cable (which often determines ampacity). If you will be working in cramped spaces and have a need for flexibility of your cable, make sure you balance this carefully with your ampacity and insulation needs. Flexible, insulated cable does exist, but it might come with a higher price tag, so make sure you prepare for that possibility.

5. Strand Count

You might have come across this factor while researching cable flexibility, but it deserves its own section for a couple of reasons. Strand count refers to the number of individual copper strands that make up the interior of the cable. This is very different from gauge, which we will explore below. Strand count has a very high effect on the flexibility of a cable, and if you are in need of particularly flexible cables with high ampacity ratings, you will most certainly want to search for high strand count cables. A word of warning, however: high strand count will increase the cost of a cable significantly, as the manufacture of high strand count cables is more complicated than lower strand counts.

6. Gauge

The gauge of a wire is the most impactful factor on a cable's ampacity. Because of this fact, though, it means you're least likely to encounter gauge when shopping for ampacity because the ampacity rating will factor in gauge as a part of the score. That said, there are still times in which knowing the gauge is incredibly important. Certain tools will have demand for specific gauges, and if your welding torch is limited to certain gauge types, you'll need to factor that in. Gauge can also be important if you are ordering cables directly from the warehouse, where ampacity ratings might not be as easily checked. Knowing the gauge of the cable, the length you'll need, and the insulation types will let you decide for yourself which cable is right for the job.

Ground Clamps

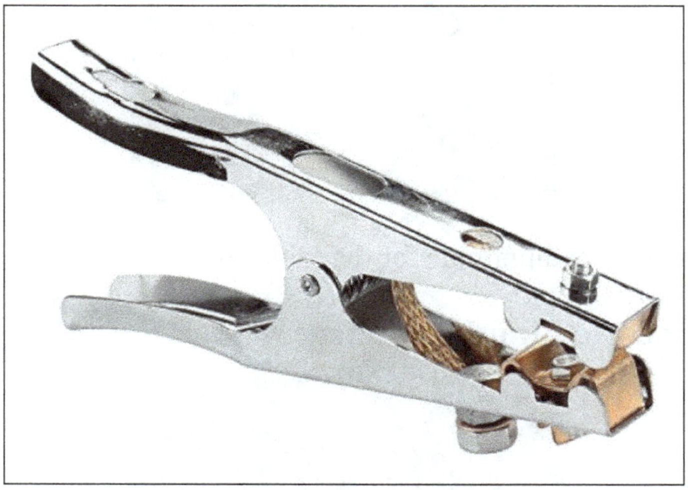

A good ground clamp is essential to produce quality welds. Without proper grounding, the circuit voltage fails to produce enough heat for proper welding, and there is the possibility of damage to the welding machine and cables. Three basic methods are used to ground a welding machine. You can fasten the ground cable to the workbench with a C-clamp, attach a spring-loaded clamp directly onto the workpiece, or bolt or tack-weld the end of the ground cable to the welding bench. The third way creates a permanent common ground.

How to Improve Ground Clamps

Step 1. Fit a Piece of Aluminum Bar Stock

We can fit some aluminum bar stock inside one half of the stamped steel spring clamp. My bar stock is 1/8 inch x 3/4 inch. The space inside the stamped steel clamp handle is about 9/16 inch. Then we can use a hacksaw to trim off the excess.

Step 2. Drill a Hole in the Aluminum

Drill a hole through the piece of aluminum for the cable bolt.

Step 3. Cut and Bend the Aluminum Piece

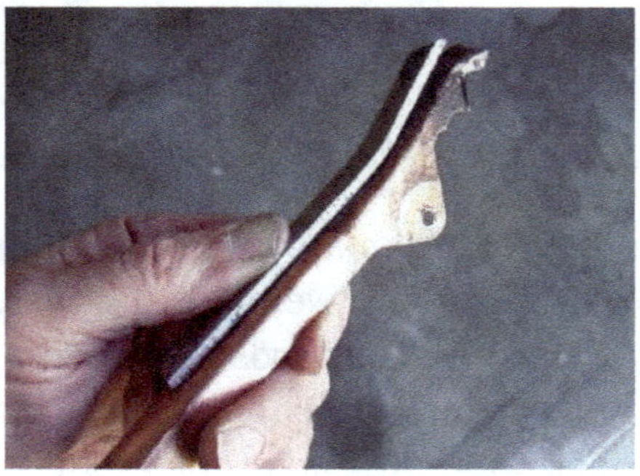

In this photo we have drilled out and removed the rivet that holds the clamp together, removed the plastic handle grip, loosened the bolt, removed the cable, cut the piece of aluminum to length, and bent it to fit the contours of the stamped piece of steel.

Step 4. Build Up the Thickness

We cut a second piece of aluminum to thicken it inside the jaws so it can clamp directly against the steel we are welding.

Step 5. Drill for a Brass Screw

We have drilled through both pieces of aluminum at the jaw and the steel of the jaw, used a countersink to prepare the aluminum for a bevel head brass screw, inserted the screw and fastened it with a brass nut. We cut the brass screw end nearly flush with the nut and peened it with a hammer to keep it from coming loose in the future.

Step 6. Inserting the Spring

The spring is quite strong. We did not want to wrestle it back into the handle halves while trying to insert a screw to replace the rivet. We decided to squeeze the spring in my vise and tie it in its squeezed position with some sturdy wire.

Step 7. Assembling the Two Halves of the Clamp

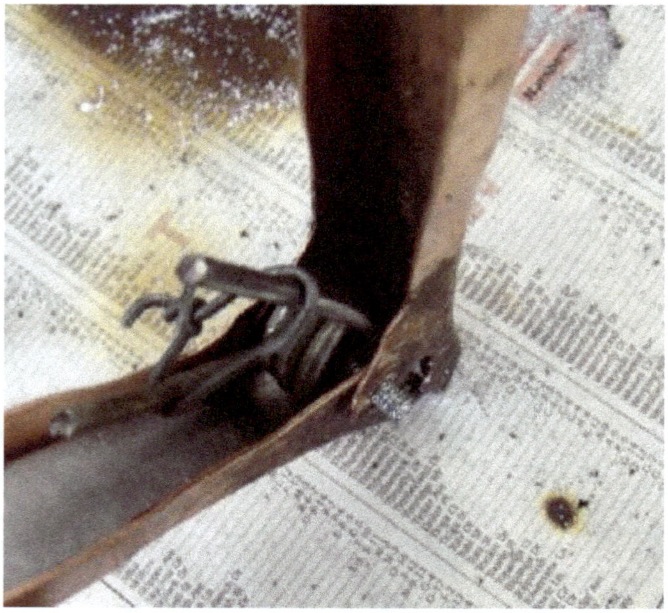

Line up the holes in the two halves of the clamp. Slide the spring into the clamp and insert a screw to replace the rivet so that the screw also passes through the spring. Put a locking nut onto the end of the screw. Make certain one of the plastic handle grips is already on the ground cable. Bolt the cable to the clamp. The bolt now passes through both the steel handle and the hole you drilled in the aluminum.

Step 8. Finish Reassembly

Cut the wire retainer on the clamp spring. Remove the wire with a pair of pliers. Slide the plastic grips onto the spring clamp.

Step 9. Grind Away Steel on the Improved Jaw

Grind away raised portions on the jaw to which aluminum has been added. Here you see we have ground away some of the steel. We may yet grind away the remaining peaks.

Step 10. Getting the Maximum Contact Area

In the photo you can see we did grind away the remaining peaks on the improved side of the clamp (bottom half in the photo). We also filed the screwhead so it does not rise above the aluminum in the jaw at all. And, We have inserted a piece of scrap steel between the jaw and the piece to which the clamp is attached. This makes a pressure point in the middle of the clamp jaw's length that pulls the flat aluminum added to the jaw firmly against the work piece for a broad area of electrical contact. If one were welding on a round tube, it would perform the same function without an added piece of scrap.

CHAPTER 3 Types of Weld Joints

A welding joint is the point where two or more pieces of plastic or metal are joined together. It is of several types such as butt joint, corner joint, lap joint, edge joint and tee joint. This chapter examines such types of weld joints and the ways to perform these as well as the techniques of squaring, clamping and tacking joints and weld.

Butt Weld

Butt welding is a commonly used technique in welding that can either be automated or done by hand on steel pieces. Butt welding can also be done with brazing for copper pieces. It is used to attach two pieces of metal together such as pipe, framework in factories, and also flanges. A flange is something that either is internal or external that provided to strengthen a piece of material. In factories butt welding has shown how economical it can be for companies to use when building things out of metal. This is because if they wanted to make something out of metal without welding it together they would have to bend everything and reinforce the structure which costs more than welding the two pieces together. Butt welding is accomplished by heating up two pieces of metal, or applying pressure, or doing both of those. Penetration while welding the metal is important to maintain and with thin pieces of metal this is possible however, with thick pieces edge preparation may have to be done to prepare the metal. Full penetration butt welds are made when they are in the within the parent(bigger, stronger) metal. In butt welding the strongest welds will have the least amount of imperfections. To achieve this the heat input is controlled, which decreases the size of the weld. In commercial welding when this is done it also reduces cost but in order to maintain the strength of the weld double butt welds will be used. In butt welding there are two types used to achieve the specific welds and then there are also a variety of joints considered to be butt joints.

Butt welding is best performed with MIG or TIG welding applications due to their natural ability to connect two pieces of metal together. Using different types of welding electrodes for the welder will determine the properties of the weld such as its resistance against corrosion and strength. Electrodes conduct current through the metal being welded in order join the two pieces. The metal determines the type of welding that is required. The electrodes are either heavily or lightly coated. For the heavily coated electrodes are commonly used in structural welding because they are much stronger and corrosion resistant. The lightly coated electrodes are not as structurally sound. Butt welding is performed with the Arc, TIG, or MIG welder held at a slight angle the weld if the weld is laying flat in order to achieve the least amount of porosity in the weld and also to increase the weld's strength. Fillet welding make up about 80 percent of the connection despite being weaker that butt welds. The reason it is used more often is because fillet welds offer more room for error with much larger tolerances. Fillet welding is not a type of butt weld despite its similarities.

Types of Butt Welding

Flash

Flash butt welding is used with machinery and connects multiple pieces of metal together that are miss matched in size and shape. These different sizings can oftentimes cause for breaks in welding process. High voltage current is applied in order to connect the metal pieces together by applying it to both the components known as flashing in order to join them together.

Resistance

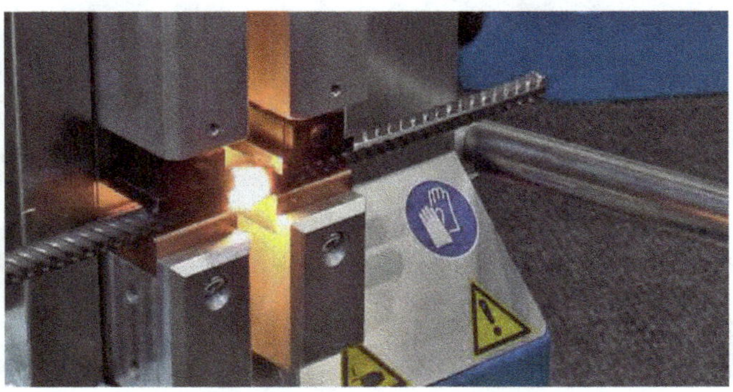

This weld joins the two pieces of metal together by heat that comes from the pressure due to the metals being held together at a preset force. Resistance butt welding is used on joints that are of similar shape and size and often the weld is performed in one movement unlike flash welding.

Types of Butt Joints

Single-V Butt Weld

There are many different types of butt welding joints and they all are named with their particular shape. The joint also known as a square groove weld has many different forms in order to connect pieces of metal together and are all capable of bearing loads. There are many different types of joints such as lap joints, tee joints, butt joints, and also corner joints. Lap joints are two pieces that are end-over-end and welded together whereas butt welds are put end to end and connected that way. Butt welds are connected to each other with the thickness of the parent metal. There are many different kinds of butt welds such as square, single v, double v, single bevel, double bevel, single u, double u, single j, and also a double j. Minimizing the distortions in a weld is important however doing so will minimize the chances of full penetration. In order to get full penetration double welds such as double v, double j, and double u may be used.

How to Perform a Butt Weld using Electric Arc Welding

A butt weld is the easiest and most simple joint to weld, and also very common. This is a great joint to use when practicing welding as a beginner.

Steps

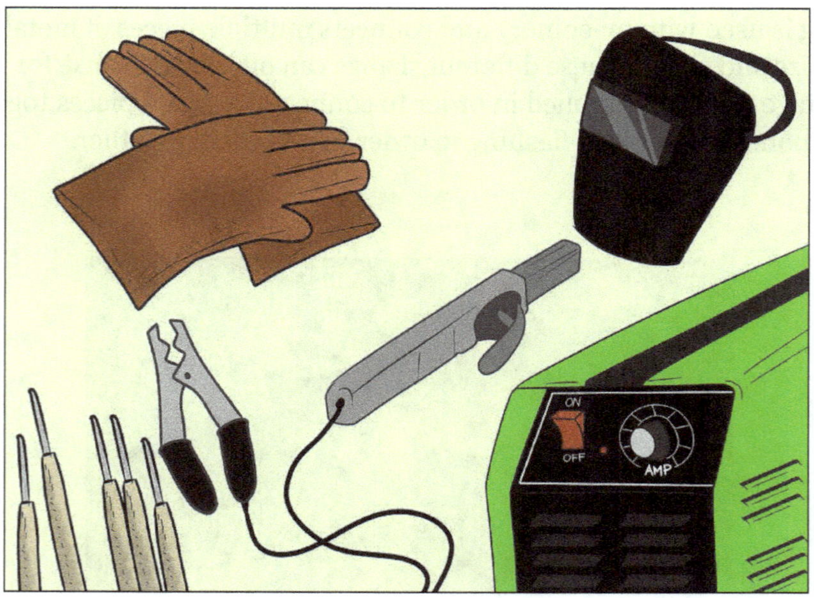

1. Obtain all necessary equipment. This should be a welding machine, electrode and workpiece clamps (and their leads), a welding helmet darker than shade 10, welding gloves, and appropriate safety clothing.

2. Prepare the metal to be welded. This includes grinding down rough edges and cleaning the areas to be welded.

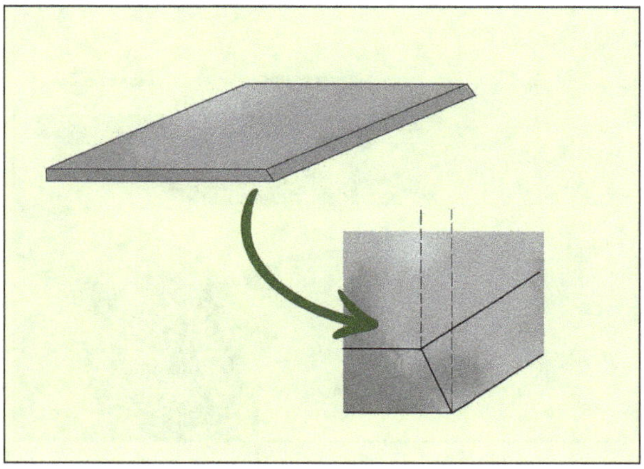

3. Bevel the edge of the metal if it is thicker than $\frac{1}{4}$ inch (0.6 cm). Beveling allows for better penetration of the root pass and subsequent passes. Beveling can be done with an oxy-fuel torch or a plasma arc cutter, but isn't necessary on thinner metal.

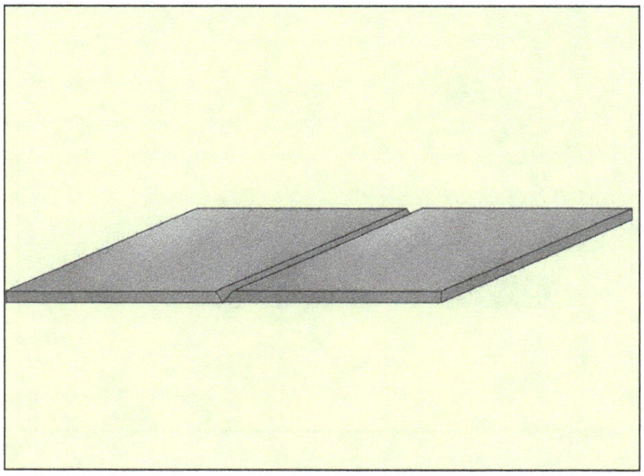

4. Align your metal to make sure the edges line up well. They should be smooth and align cleanly.

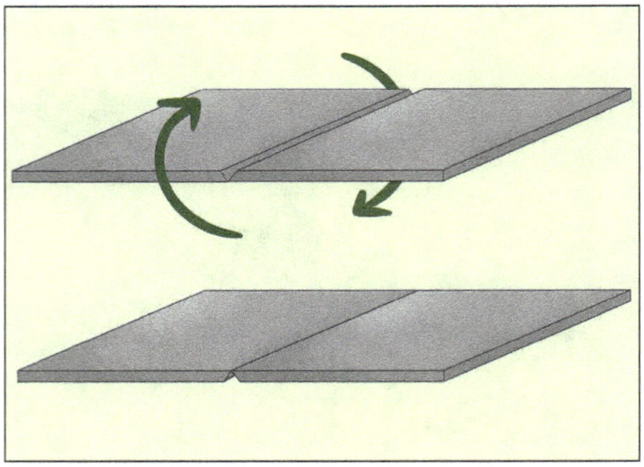

5. Turn your pieces over. This should be the flat side if on or more pieces are beveled, or the side you don't want to start welding.

6. Separate the pieces a little bit and set the amperage on your machine about 10 percent higher than the level you intend to use welding the metal. So, if you are going to use 100 amps to make your welds, set your amperage at 110 amps.

7. Make tack welds. These will hold the metal together and prevent it from warping or bending inward when the weld is finished. To make a tack weld, strike an arc and let it sit for a few seconds. A few tack welds are usually needed and you should be able to break them with a hammer or wrench.

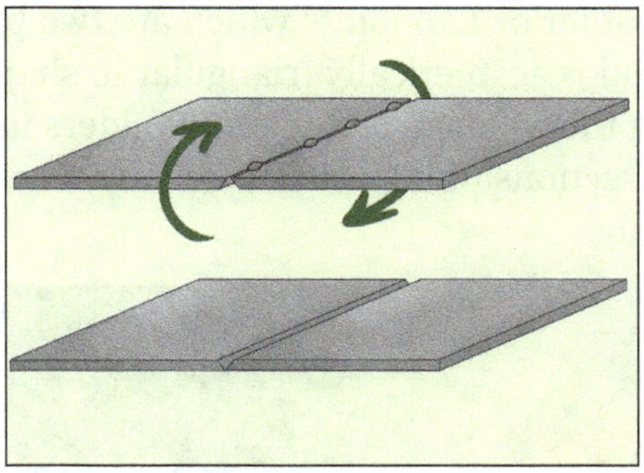

8. Flip your metal over to be welded.

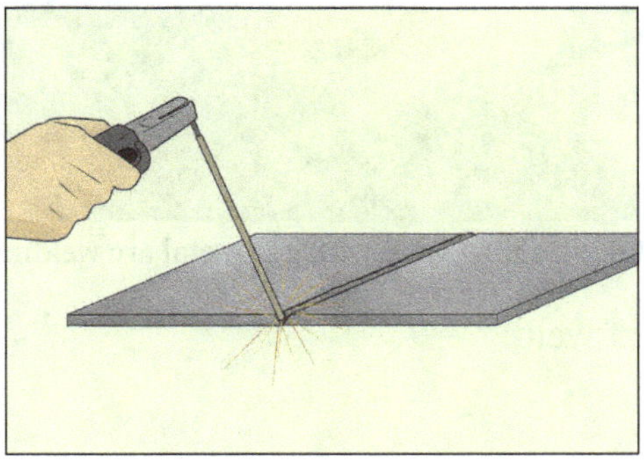

9. Strike an arc and create your root pass. This is going to be the first and deepest pass on your weld, and if the metal is thick enough, the only pass you'll need. If you beveled the steel start at the bottom for your root pass. You need to ensure the root pass penetrates deeply enough, and for this reason 6010 electrodes are frequently used for this purpose.

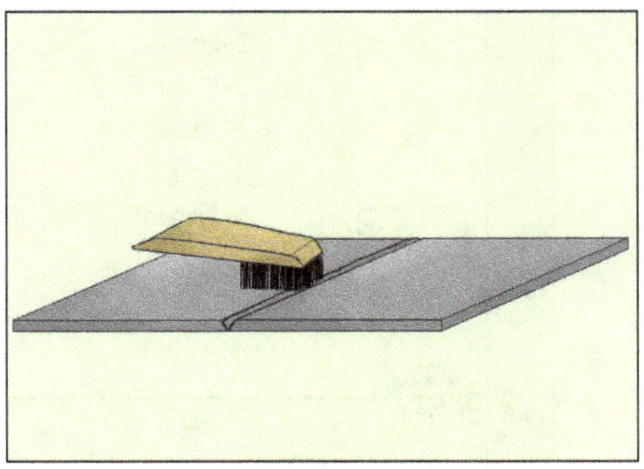

10. Clean the weld with a hammer and wire brush and make subsequent passes if needed. These passes should strengthen the weld and fill it in. Make sure to clean each pass before you start a new one.

Fillet Weld

Fillet welding refers to the process of joining two pieces of metal together whether they be perpendicular or at an angle. These welds are commonly referred to as Tee joints which are two pieces of metal perpendicular to each other or Lap joints which are two pieces of metal that overlap and are welded at the edges. The weld is aesthetically triangular in shape and may have a concave, flat or convex surface depending on the welder's technique. Welders use fillet welds when connecting flanges to pipes, welding cross sections of infrastructure, and when fastening metal by bolts isn't strong enough and wear off easily.

Making a fillet weld with gas metal arc welding

There are two main types of fillet weld:

1 - Traverse Fillet weld

2 - Parallel Fillet weld

Aspects

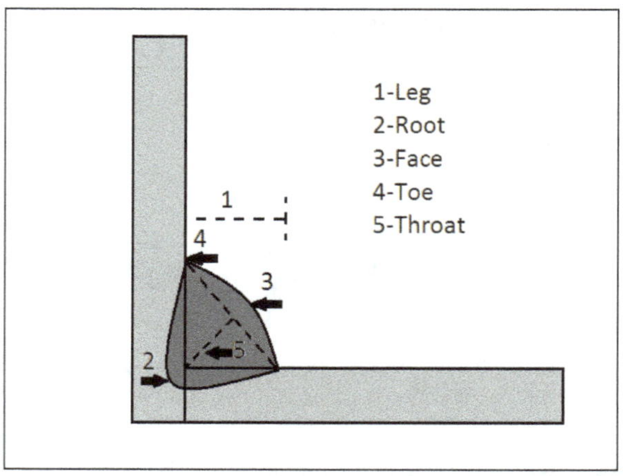

Parts of a fillet weld

There are 5 pieces to each Fillet weld known as the Root, Toe, Face, Leg and Throat. The root of the weld is the part of deepest penetration which is the opposite angle of the Hypotenuse. The toes of the weld are essentially the edges or the points of the hypotenuse. The face of the weld is the outer visual or hypotenuse that you see when looking at a fillet weld. The legs are the opposite and adjacent sides to the triangular fillet weld. The leg length is usually designated as the size of the weld. The throat of the weld is the distance from the center of the face to the root of the weld. Typically the depth of the throat should be at least as thick as the thickness of metal you are welding.

Notation

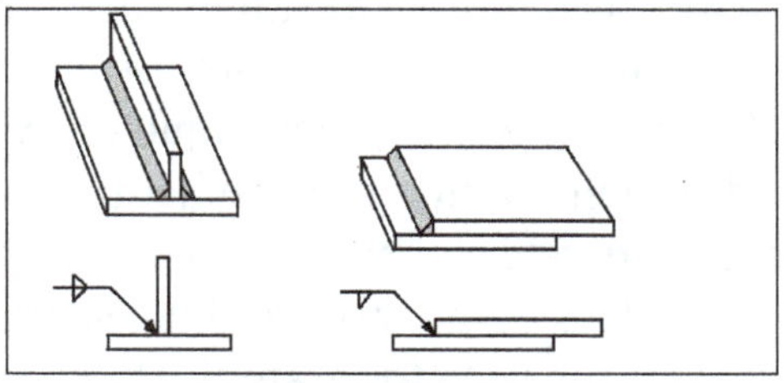

Fillet weld notation

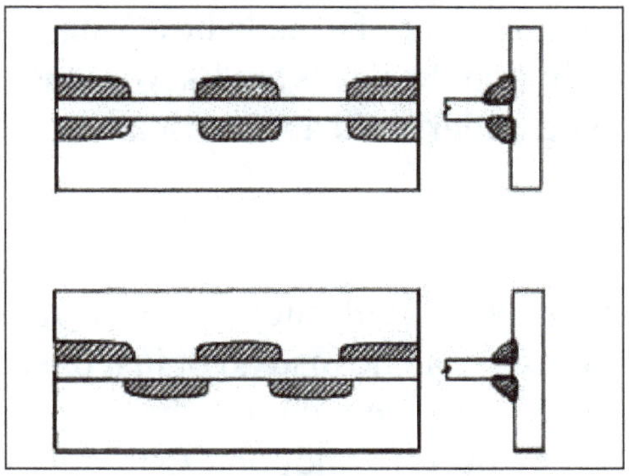

Intermittent fillet welds

Fillet welding notation is important to recognize when reading technical drawings. The use of this notation tells the welder exactly what is expected from the fabricator.

The symbol for a fillet weld is in the shape of a triangle. This triangle will lie either below a flat line or above it with an arrow coming off of the flat line pointing to a joint. The flat line is called "reference line". The side on which the triangle symbol is placed is important because it gives an indication on which side of the joint is to be intersected by the weld. It is recognized that there are two different approaches in the global market to designate the arrow side and other side on drawings; a description of the two approaches is contained in International Standard ISO 2553, they are called "A-System" (which is more commonly used in Europe) and "B-System" (which is basically the ANSI/AWS system used in the US). In "A-System" two parallel lines are used as reference line: one is a continuous line, the other is a dashed line. In the "B-System", there is only one reference line, which is a continuous line. If there is a single reference line (B-System) and the triangle is positioned below the line, then the weld is going to be on the arrow side. If there is a single reference line ("B-System") and the triangle is positioned above the line, then the weld is going to be on the opposite side of the arrow.

When you find an arrow pointing to a joint with two triangles, one sitting below and one sitting above the line even with each other, then there is intended to be a fillet weld on the arrow side of the joint as well as the opposite side of the joint.

If the weld is to be continuous around a piece of metal such as a pipe or square, then a small circle will be around the point where the flat line and arrow pointing to the joint are connected.

Manufacturers also include the strength that the weld must be. This is indicated by a letter and number combination just before the flat line. Examples of this are "E70" meaning the arc electrode must have a tensile strength of 70,000 pounds-force per square inch (480,000 kPa; 4,900 kgf/cm²).

There are also symbols that describe the aesthetics of the weld. A gentle curve pointing away from the hypotenuse means a concave weld is required, a straight line parallel with the hypotenuse calls for a flat faced weld, and a gentle curve towards the hypotenuse calls for a convex weld. The surface of the weld can be manipulated either by welding technique or by use of machining or grinding tools after the weld is completed. When reading a manufacturers technical drawings, you might also come across weld dimensions. The weld can be sized in many different ways such as the length of the weld, the measurements of the legs of the weld, and the spaces between welds. Along with a triangle, there will usually be a size for the weld for example (1/8"x3/8") to the left of the triangle. This means that the vertical leg of the weld is to be 1/8" whereas the horizontal leg is to 3/8". To the right of the triangle, there will be a measurement of exactly how long the weld is supposed to be. If the measurements of the drawing are in mm the welds are likewise measured in mm. For example, the weld would be 3 x 10, the mm being understood automatically.

Intermittent Fillet Welds

An intermittent fillet weld is one that is not continuous across a joint. These welds are portrayed as a set of two numbers to the right of the triangle instead of just one. The first number as mentioned earlier refers to the length of the weld. The second number, separated from the first by a "-", refers to the pitch. The pitch is a measurement from midpoint to midpoint of the intermittent welds. Intermittent welding is used when either a continuous weld is not necessary, or when a continuous weld threatens the joint by warping. In some cases intermittent welds are staggered on both sides of the joint. In this case, the notation of the two triangles aren't directly on top of each other. Instead, the side of the joint to receive the first weld will have a triangle further to the left than the following side's triangle notation. As an end result of alternating intermittent fillet welds at each side, the space between welds on one side of the joint will be the midpoint of the opposite side's weld.

How to Perform a Fillet Weld

The most common type of welded joint you will encounter is known as a T-fillet weld. This type of joint crops up everywhere, because it is easy to design into a structure. If you want to be good at welding, the techniques involved in producing this joint in all materials and welding positions must be mastered.

1. Horizontal & Vertical Position

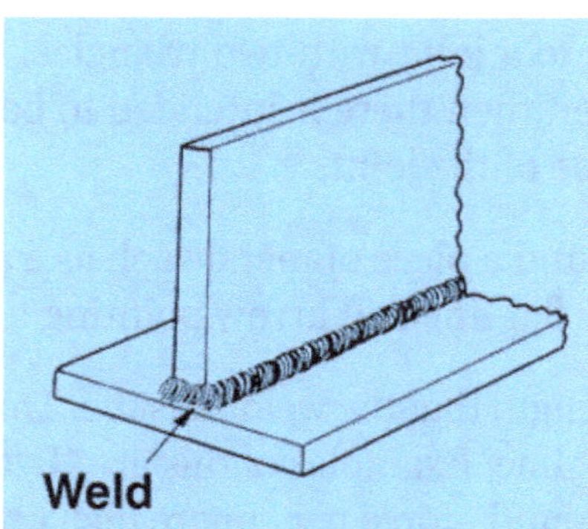

*This image shows the line of the joint in the horizontal/vertical welding position (PB).

2. Joint Preparation

Good welding is all about preparing the parent metal material before you weld.

Essentially, you need to weld metal that is not heavily rusted, covered in paint, grease or oil. The joint faces, where the metals meet, must be touching and not bent or twisted with large gapping. Good welding is 95% effort up front. If you can't prepare your joint correctly, don't weld it.

To remove paint and rust, use an angle grinder with a correctly mounted grinding wheel or sanding pad. Be gentle, as excessive grinding can leave deep grinding marks and reduce the metal thickness considerably.

When grinding, DO NOT forget to wear the correct PPE, namely eye protection correctly rated for a grinding application and some protective gloves.

Having removed rust or paint, spend some time on ensuring that the edges of the parent materials meet, wherever possible try to avoid twisted or bent metals and ensure the two pieces touch.

It might be an idea to use degreasing agent to remove any remnants of oil or grease. It only takes seconds to wipe the joint faces with a degreaser.

There are many safe degreasing agents available commercially to purchase. Try to avoid chemically based solutions and definitely do not use paint thinners. For advice, speak to Foster Industrial.

3. What do you need to Start Welding

- 2 x pieces of 6mm low carbon steel approximately 150mm x 50mm. Remove the mill scale using a sanding pad and angle grinder from the joint faces.
- A G-clamp.
- An engineer's square.
- A hammer.
- Your welding machine and PPE.
- Diameter 1.0mm low carbon steel filler wire.
- A shielding gas mixture of argon plus either oxygen or carbon dioxide.

4. Tack Welding

Welding generates huge amounts of heat. This heating and cooling has the effect of bending or twisting the joint during welding. This known as distortion. We cannot stop this from happening, but we can control and limit it's effect.

The most common way, is to strategically place some small welds along the joint to hold the joint in alignment during the welding process. These are known as tack welds. It is normal to tack weld every welded joint you attempt. Note: The thinner your parent metal is the more tack welds you will need.

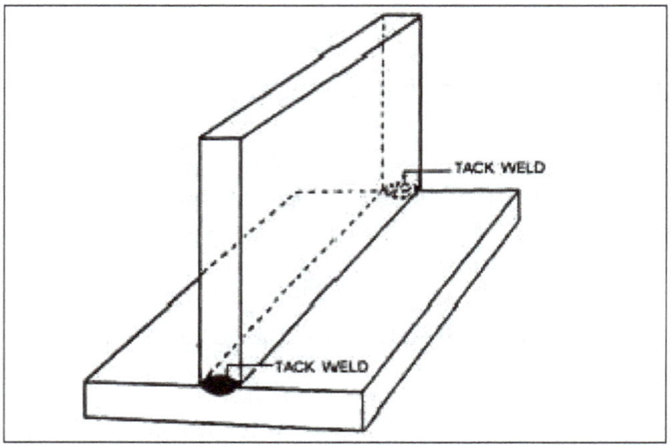

- Set up and clamp the two pieces into a T shape.
- Tack weld the two ends of the T-joint.
- Tap the joint to 90° and check using the square that the angle is correct.
- Use the leftwards technique to weld the joint.

Look at the angle of the welding torch. This should be 70° to 80° and note the direction of travel. The leftwards technique means that the torch is pushed along the joint during welding or the gas pocket is always ahead of the welding wire. The direction of travel is right to left. For left handed people, the direction of travel will be left to right with the gas pocket also ahead of the welding wire.

Refer to your completed chart from Lesson 3 and find machine settings that you think will allow you to produce a T-fillet joint. To assist, you will probably need to be setting your volts to somewhere between 21V – 23V. Adjust your wire feed so that you are in dip transfer & hear the 'bacon frying' noise.

5. How a Good T-fillet Weld Must Look

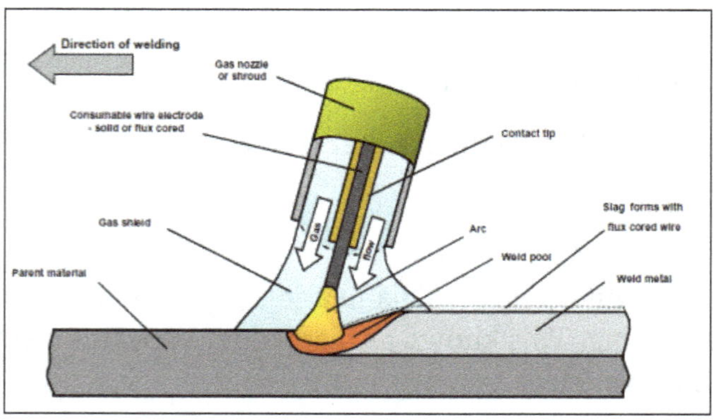

After welding use an angle grinder and sanding pad to remove localised spatter.

If you need to re-paint the welded area ensure that the joint and adjacent materials have cooled down completely.

Lap Joint

A lap joint is the joint between two pieces of metal in which the edges or ends are overlapped and fastened together to produce a continuous or flush surface.

Several types of lap joint techniques exist and are employed depending on the type of materials, shape, and thickness of the parts or the application of the jointed object. Lap joints are also applicable in wood and plastics.

Full-lap and half-lap joints are the most commonly used joints. In the full-lap technique, no material is removed from any of the parts and the resulting joint is the combined thickness of the two. In the half-lap joint, some material is removed from each of the members and the joint is the thickness of the thickest part. If the two members have similar thicknesses, only half is removed from each.

A lap joint may be temporary or permanent. Temporary joints may be used in most engines and other forms of machinery where replacements are required or when two parts of the machine are joined together to achieve a particular task. This uses brackets and fastening components such as bolts, screws, nails and rivets. Permanent joints are formed using brazing, welding, soldering, flame joints and adhesives.

Advantages of lap joints include:

- Easy to prepare (does not require cut faces to be parallel or perfectly flat)
- Can be formed between two dissimilar metals, such as aluminum and copper
- Accommodates different thicknesses (thinner piece must be welded on top)
- Thin material such as diaphragms and foils can be joined

Disadvantages include:

- Some instances of lower tensile strength
- Less rigid than the base materials since the weld may act as a pivot
- Overlaps may be undesirable for mechanical or aesthetic reasons.

- Micro-cracks and cavity defects may occur if wrong welding speed is used.
- Corrosion and fatigue cracking may occur on the shielded areas due to moisture retention

Lap joints are widely used in woodworking, plastics and metals. They are used to enable the fabrication of regularly and irregularly shaped parts, sheets and also when longer parts are required. Typical applications include aircraft fuselages and the advanced structural frames for cars and motorcycles.

Corner Joint

Corner joints are used to join two members located approximately at right angles to each other in the form of an L. The fillet weld corner joint is used in the construction of boxes, box frames, and similar fabrications. The closed corner joint is used on lighter sheets when high strength is not required at the joint. In making the joint by oxyacetylene welding, the overlapping edge is melted down, and little or no filler metal is added. When the closed joint is used for heavy sections, the lapped plate is V-beveled or U-grooved to permit penetration to the root of the joint. The open corner joint is used on heavier sheets and plates. The two edges are melted down, and filler metal is added to fill up the corner. Corner joints on heavy plates are welded from both sides.

Edge Joint

Edge joints are used to join two or more parallel or nearly parallel members. Edge joints are not very strong, and are used to join edges of sheet metal, reinforcing plates in flanges of I-beams, and for edges of angles. Two parallel plates are joined together, as shown in figure below. On heavy plates, sufficient filler metal is added to fuse or melt each plate edge completely and to reinforce the joint. Light sheets are welded as shown in figure below. No preparation is necessary other than to clean the edges and tack weld them in position. The edges are fused together so no filler metal is required. The heavy plate joint, as shown in figure below, requires that the edges be beveled to secure good penetration and fusion of the side walls. Filler metal is used in this joint.

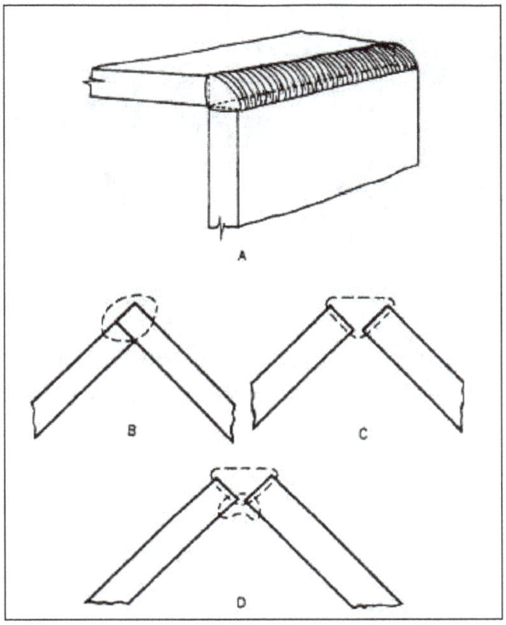

Corner joints for sheets and plates.

How to Square, Clamp and Tack your Joints and Weld

Cutting and Grinding your Parts

In the Tools and Materials section of this course, we go over how to set up an angle grinder for use with a cutting wheel and flap wheel grinding disk. Cutting steel bar, rod, and tube is most affordably done using an angle grinder and cut off wheel. Making nearly straight cuts on flat bar with a cut-off wheel can be acheived by rotating the guard on the grinder, and positioning cut off wheel straight down perpendicular to the bar. This method can be imprecise and lead to more grinding work later.

Clamps and Tacking

Welding clamps are used to hold work together before you tack your parts together. Clamps come in more sizes and shapes than you can even dream of. My favorite and most commonly used work-holding clamps for welding are:

- Welding magnets
- Deep-reach vice grips
- Squaring two-axis welding clamp

Be sure to check your square as you tack pieces together with a steel square tool. If your tacks are not keeping your material square, you should be able to bend and twist (or even bang and hammer) them back to square before completing your weld beads.

We tack our material to minimize heat distortion on our material. Rapid expansion and contraction will cause your material to shrink toward your weld bead, so alternating sides as you tack can help reduce this deformity.

Tee Joints

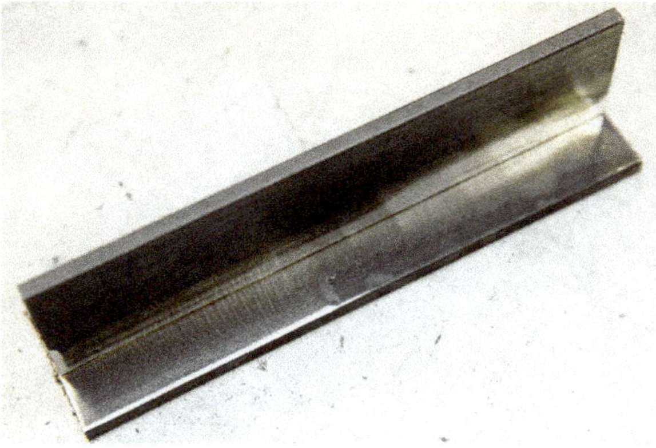

To learn the Tee joint, we first need to introduce the concept of fillet welding. Fillet welding refers to the process of joining two pieces of metal together whether they be perpendicular or at an angle. To perform a fillet weld, we change the way we position our welding torch against the material, so that we penetrate into the corner of our joint at an angle that will be well shielded and ensure strength.

The weld bead will not penetrate all the way through the other side of your base material like we see when we are welding on the sheet material from the Beginning to Weld Lesson.

The tee joint is when two pieces of metal are perpendicular to each other at a 90 degree intersection. Prepare your metal for welding by grinding off the mill scale finish. You only need to grind off the mill finish where you intend to weld, remember to keep considering how to minimize your grind time.

For a clean and close fitting weld, I recommend using magnets that will hold your material square to one another while you are making your first few tacks. Be sure to check your square between tacks using a slide rule or metal square.

Start by tacking each end, placing a tack weld on either side of the tee. Complete a tack in the center, before tacking the opposite side of the tacks you already placed in the center. Finish by tacking the back side of your previous center tack.

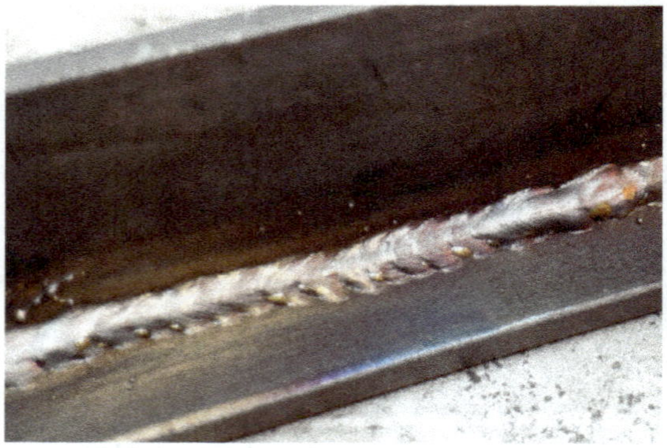

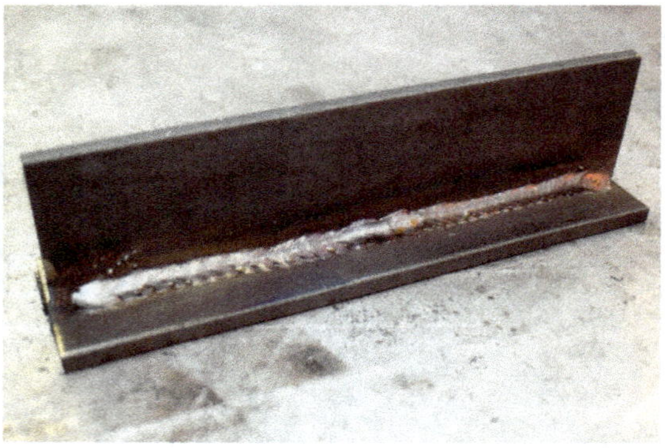

A fillet weld motion is not quite the same the chain of loops like we did for our butt weld, instead it is more like a rounded zig-zag shape, where we grab and melt the edges of the base material as we run the weld puddle across our joint. Run beads in sections along your joint, alternating side to side and back to front, working your way towards the middle. A nice fillet weld will form a neat even triangle shape between the pieces of perpendicular material.

Corner Joint

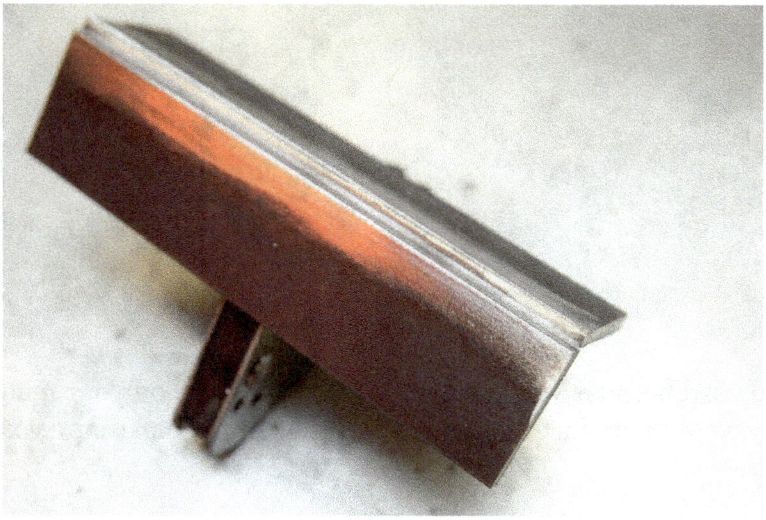

A corner weld is a type of joint that is between two metal parts and is located at right angles to one another in the form of an L.

To complete this joint, begin by tacking the outside edges, then make the same curved zig-zag weaving motion we made for our filleted tee weld.

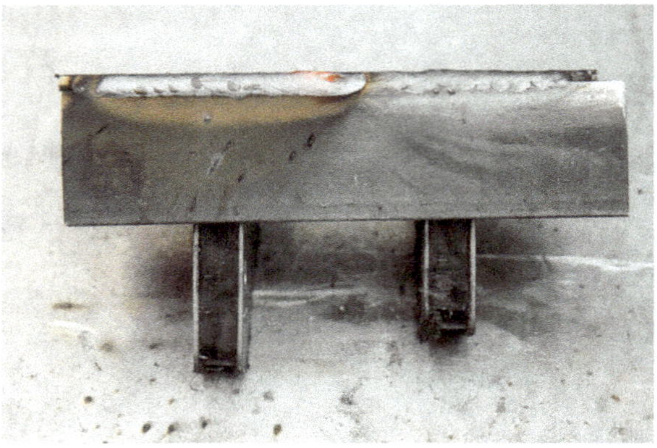

Corner joints can be hard because you can't often rest your hand on your material to steady your torch hand. You may want to practice a dry run along the joint to make sure you're going to be able to weld comfortably,

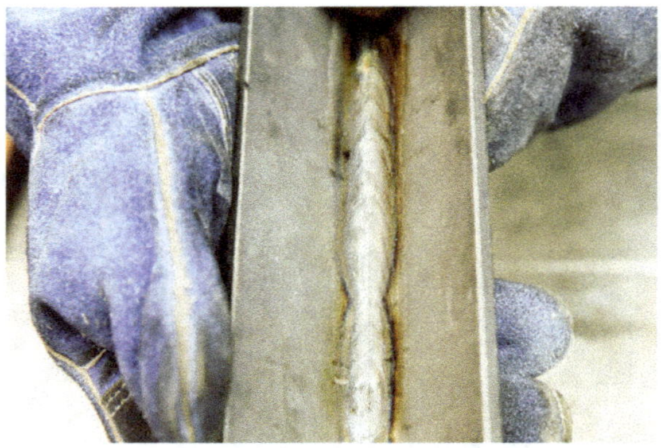

Lap Joint

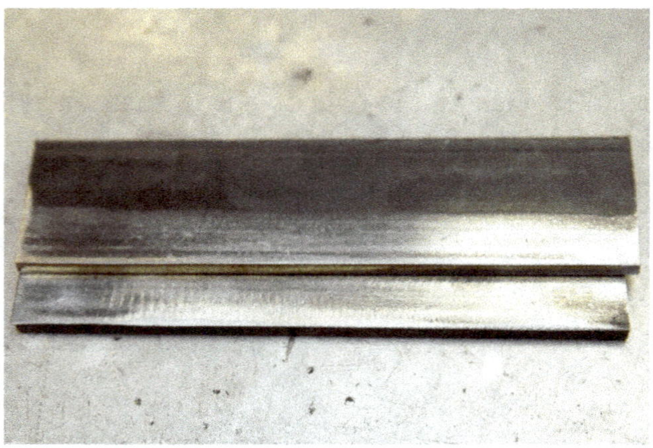

Lap joints will need to be completed when your project has one piece of material nested on top of another sheet or bar, with a large amount of surface area being mated. This joint is very similar to the tee joint, and we will need to perform a fillet weld to merge the material into one continuous piece of steel. Lap joints often necessitate slightly more voltage and a slightly decreased wire feed because there are no gaps in your material, and you are essentially welding onto a heat sync. The bottom layer of welding material is absorbing the heat energy of the top piece of material, dissipating the heat quickly.

Prepare the edges of your material to be welded. Begin by tacking the pieces together. Instead of making a small cursive E shape, place your consumable wire electrode on your top piece of material, squeeze the trigger, and gently drag the gun toward the bottom piece of material. You should be able to see the edge of the top material ever so slightly melt away through your welding hood as you complete this motion.

Complete your lap joint by first running beads along the outer edges, and work your way towards the middle. The angle of the gun should go into the corner at 45 degrees. The weld pool should visibly 'grab' the edge of the steel on your top piece of material. Remember that in a tee joint, your are not making small loop motions while you weld, but instead a slightly rounded zig-zag.

Edge Joint

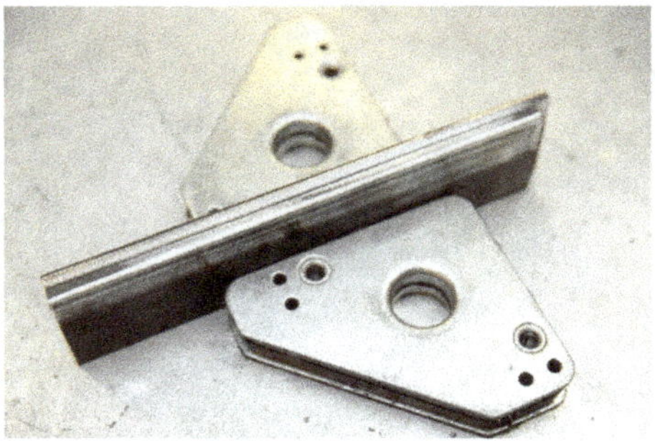

An edge joint is similar to the lap weld, in that we will be grabbing the corners of our sheet or bar material as we weld the edges together, melting them into our weld pool. Position the torch parallel to the edge and use the same zig-zagging motion as the torch moves along the path of the weld.

The heat affected zone of an edge weld is greater than any other joint we have gone over. Because the heat is spreading into two parallel pieces of material, with the same amount of heat being applied to both pieces, the material can burn out quickly. To avoid a gap in your weld or material, edge welds will often need a lowered voltage setting.

Begin by tacking the two ends, then place two tacks between the two edges, equidistant apart - four tack welds in total.

Run a bead on one end, then the other end, completing the middle bead last. By welding the ends first, we minimize heat distortion and potential burn-out.

If your material is very thin, consider using a small lap weld instead of an edge weld; as you perform this lap weld, you pick up the edges of both workpieces instead of just the top piece. You'll have an easier time manipulating your puddle and a slightly smaller heat affected zone.

CHAPTER 4

Shielded Metal Arc Weld: Problems and Solutions

Some of the problems associated with shielded metal arc welding are weld spatter, poor fusion, cracking, shallow distortion, etc. Besides the above, there can arise distortions, warping of thin plates and brittle welds. The strategies for the control and prevention of distortion, warping, corrosion, undercutting, spattering, etc. have been addressed in this chapter in great detail.

Weld Distortion

Because welding involves highly localised heating of joint edges to fuse the material, non-uniform stresses are set up in the component because of expansion and contraction of the heated material. Initially, compressive stresses are created in the surrounding cold parent metal when the weld pool is formed due to the thermal expansion of the hot metal (heat affected zone) adjacent to the weld pool. However, tensile stresses occur on cooling when the contraction of the weld metal and the immediate heat affected zone is resisted by the bulk of the cold parent metal.

The magnitude of thermal stresses induced into the material can be seen by the volume change in the weld area on solidification and subsequent cooling to room temperature. For example, when welding CMn steel, the molten weld metal volume will be reduced by approximately 3% on solidification and the volume of the solidified weld metal/heat affected zone (HAZ) will be reduced by a further 7% as its temperature falls from the melting point of steel to room temperature.

If the stresses generated from thermal expansion/contraction exceed the yield strength of the parent metal, localised plastic deformation of the metal occurs. Plastic deformation causes a permanent reduction in the component dimensions and distorts the structure.

Welding Distortion can be classified into following group:

1. Longitudinal shrinkage
2. Transverse shrinkage
3. Angular distortion
4. Bowing

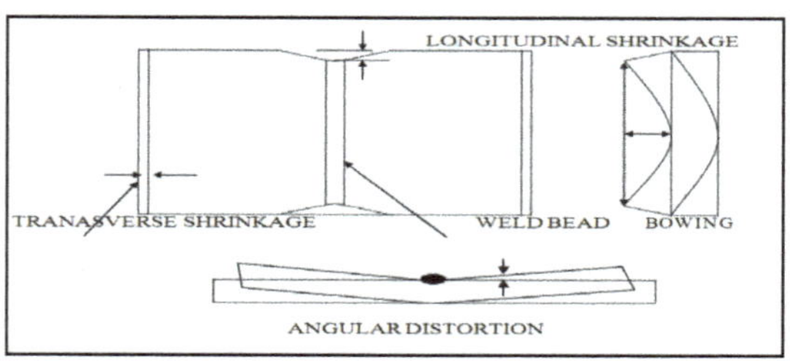

Type of distortion in welding of free plates

Longitudinal Shrinkage

When the length of the welded plate is shortened along the longitudinal axis or weld shrinkage parallel to the weld, then it is called as longitudinal shrinkage. When a weld is deposited lengthwise on a light narrow and perfectly flat strip of metal that is neither compiled nor held in any way, the strip will tend to bow upward in the direction of the head. This is occurring due to the longitudinal contraction of the weld metal as it cools. The value of longitudinal contraction is maximum along the weld centre line and decreases towards the edges. Minimum distortion occurs due to High current and large size electrode. Also low current with small size electrode causes minimum distortion. But maximum distortion occur in-between these two situations.

Longitudinal Distortion Depends Upon

i) Contraction forces.

ii) Distance between the centroids of weld and section.

iii) Stiffness of the section being welded.

Transverse Shrinkage

When two plates being butt welded together are neither too heavy nor held together, and are thus free to move, they will be drawn closer together by the contraction of the weld metal. This is called transverse contraction. Transverse contraction exists all along the weld length and it depends upon the permanent contraction of elements in the weld zone. On the other hand when the welded plate is shortened in a direction transverse to the bend then it is called as transverse shrinkage. Transverse contraction exists all along the weld length and it depends upon the contraction of elements in the welded zone.

Following Points can be used for Prevented the Transverse Contraction

i) Proper tack welding,

ii) Wedge placing between the plates,

iii) Scarpereding the plates (before welding) to provide allowance (about 1/100mm. of weld) for contraction,

iv) Increasing the arc travel speed.

To analyse the shrinkage mechanism, let us consider the heating of a material held between two immovable faces upon heating the material would try to expand to a new length.

Where,

ΔL is the product of coefficient of linear expansion, temperature difference and initial length. But as it cannot lengthen, it must up set by an amount represented by L on cooling it shrinks to a dimension (L - ΔL) approx. The expansion on heating is given by

$$\Delta L = \alpha (T2 - T1) L$$

Where,

α = Coefficient of linear expansion

L = Initial length

T_2 = Final temperature

T_1 = Initial temperature

A weld upon cooling in a real structure however is not rigidly restrained but is shrinking against an elastic structure. Further the surrounding material has expanded against the weld and also shrinking.

Angular Distortion

When two bevelled plates are welded, it is found that the plates are pulled out of line with each other. Since the opening at the top of the single Vee groove is greater than at the bottom, a greater portion of the weld metal is deposited there, and thus the drawing or pulling is greatest on that side of the joint. Angular contraction is related to the shape and size of the cooling weld metal zone and the stiffness of the remaining unused part. Double groove joints tend to minimize angular distortion because the contraction effects of the two sides, i.e., top and bottom of the plate, get cancelled with each other.

Bowing Distortion

Bowing is bending of weld structures along the length of the weld. This type of distortion is also called longitudinal distortion. Hence after welding the plate takes the shape of a bow i.e. when the longitudinal shrinkage does not occur along the neutral axis of the member it causes bowing. This is a result of non-uniform shrinkage through the thickness of the plate (i.e. due to difference in shrinkage on the face side and the root side of the plate). Bowing also depends upon the location of horizontal neutral axis and the location of the weld lead with respect to horizontal neutral axis id the case of welded built up section.

How to Control Distortion in Welding

To prevent or minimize weld distortion, methods must be used both in design and during welding to overcome the effects of the heating and cooling cycle. Shrinkage cannot be prevented, but it can be controlled. Several ways can be used to minimize distortion caused by shrinkage:

1. Do not Overweld

The more metal placed in a joint, the greater the shrinkage forces. Correctly sizing a weld for the requirements of the joint not only minimizes distortion, but also saves weld metal and time. The amount of weld metal in a fillet weld can be minimized by the use of a flat or slightly convex bead, and in a butt joint by proper edge preparation and fitup. The excess weld metal in a highly convex bead does not increase the allowable strength in code work, but it does increase shrinkage forces.

When welding heavy plate (over 1 inch thick) bevelling or even double bevelling can save a substantial amount of weld metal which translates into much less distortion automatically.

In general, if distortion is not a problem, select the most economical joint. If distortion is a problem, select either a joint in which the weld stresses balance each other or a joint requiring the least amount of weld metal.

2. Use Intermittent Welding

Another way to minimize weld metal is to use intermittent rather than continuous welds where possible. For attaching stiffeners to plate, for example, intermittent welds can reduce the weld metal by as much as 75 percent yet provide the needed strength.

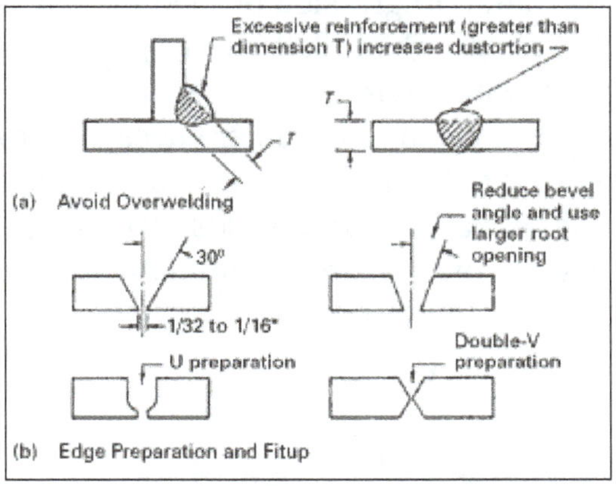

Distortion can be prevented or minimized by techniques that defeat - or use constructively - the effects of the heating and cooling cycle.

3. Use as Few Weld Passes as Possible

Fewer passes with large electrodes, are preferable to a greater number of passes with small electrodes when transverse distortion could be a problem. Shrinkage caused by each pass tends to be cumulative, thereby increasing total shrinkage when many passes are used.

4. Place Welds Near the Neutral Axis

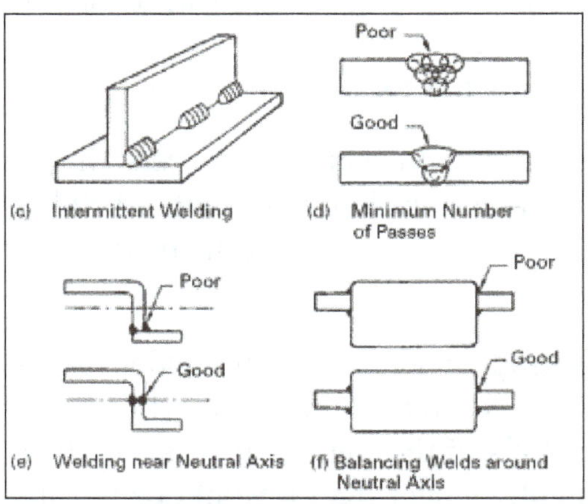

Distortion can be prevented or minimized by techniques that defeat - or use constructively - the effects of the heating and cooling cycle.

Distortion is minimized by providing a smaller leverage for the shrinkage forces to pull the plates out of alignment. Both design of the weldment and welding sequence can be used effectively to control distortion.

5. Balance Welds around the Neutral Axis

This practice, shown in figure below, offsets one shrinkage force with another to effectively minimize distortion of the weldment. Here, too, design of the assembly and proper sequence of welding are important factors.

6. Use Backstep Welding

In the backstep technique, the general progression of welding may be, say, from left to right, but each bead segment is deposited from right to left as in figure below. As each bead segment is placed, the heated edges expand, which temporarily separates the plates at B. But as the heat moves out across the plate to C, expansion along outer edges CD brings the plates back together. This separation is most pronounced as the first bead is laid. With successive beads, the plates expand less and less because of the restraint of prior welds. Backstepping may not be effective in all applications, and it cannot be used economically in automatic welding.

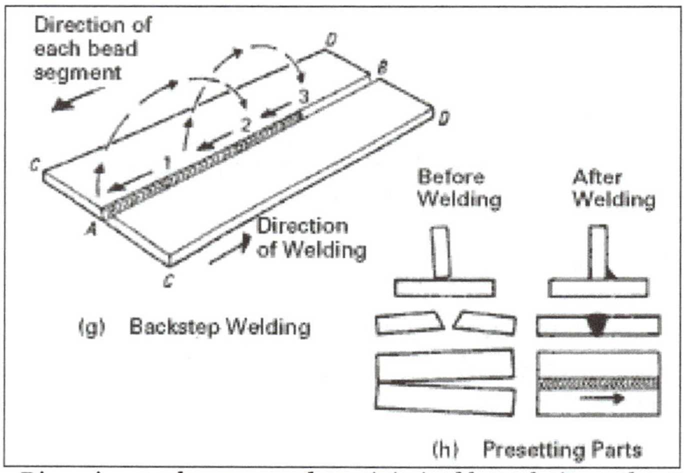

Distortion can be prevented or minimized by techniques that defeat - or use constructively - the effects of the heating and cooling cycle.

7. Anticipate the Shrinkage Forces

Presetting parts (at first glance, I thought that this was referring to overhead or vertical welding positions, which is not the case) before welding can make shrinkage perform constructive work. The required amount of preset for shrinkage to pull the plates into alignment can be determined from a few trial welds.

Prebending, presetting or prespringing the parts to be welded, above figure, is a simple example of the use of opposing mechanical forces to counteract distortion due to welding. The top of the weld groove - which will contain the bulk of the weld metal - is lengthened when the plates are preset. Thus the completed weld is slightly longer than it would be if it had been made on the flat plate. When the clamps are released after welding, the plates return to the flat shape, allowing the weld to relieve its longitudinal shrinkage stresses by shortening to a straight line. The two actions coincide, and the welded plates assume the desired flatness.

Another common practice for balancing shrinkage forces is to position identical weldments back to back, clamping them tightly together. The welds are completed on both assemblies and allowed to cool before the clamps are released. Prebending can be combined with this method by inserting wedges at suitable positions between the parts before clamping.

In heavy weldments, particularly, the rigidity of the members and their arrangement relative to each other may provide the balancing forces needed. If these natural balancing forces are not present, it is necessary to use other means to counteract the shrinkage forces in the weld metal. This can be accomplished by balancing one shrinkage force against another or by creating an opposing force through the fixturing. The opposing forces may be: other shrinkage forces; restraining forces imposed by clamps, jigs, or fixtures; restraining forces arising from the arrangement of members in the assembly; or the force from the sag in a member due to gravity.

8. Plan the Welding Sequence

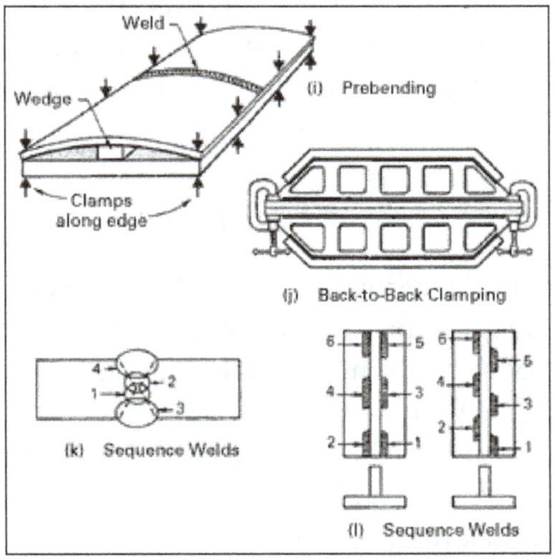

Distortion can be prevented or minimized by techniques that defeat - or use constructively - the effects of the heating and cooling cycle.

A well-planned welding sequence involves placing weld metal at different points of the assembly so that, as the structure shrinks in one place, it counteracts the shrinkage forces of welds already made. An example of this is welding alternately on both sides of the neutral axis in making a complete joint penetration groove weld in a butt joint, as in Figure above. Another example, in a fillet weld, consists of making intermittent welds according to the sequences shown in Figure above. In these examples, the shrinkage in weld No. 1 is balanced by the shrinkage in weld No. 2.

Clamps, jigs, and fixtures that lock parts into a desired position and hold them until welding is finished are probably the most widely used means for controlling distortion in small assemblies or components. It was mentioned earlier in this section that the restraining force provided by clamps increases internal stresses in the weldment until the yield point of the weld metal is reached. For typical welds on low-carbon plate, this stress level would approximate 45,000 psi. One might expect this stress to cause considerable movement or distortion after the welded part is removed from the jig or clamps. This does not occur, however, since the strain (unit contraction) from this stress is very low compared to the amount of movement that would occur if no restraint were used during welding.

9. Remove Shrinkage Forces after Welding

Peening is one way to counteract the shrinkage forces of a weld bead as it cools. Essentially, peening the bead stretches it and makes it thinner, thus relieving (by plastic deformation) the stresses induced by contraction as the metal cools. But this method must be used with care. For example, a root bead should never be peened, because of the danger of either concealing a crack or causing one. Generally, peening is not permitted on the final pass, because of the possibility of covering a crack and interfering with inspection, and because of the undesirable work-hardening effect. Thus, the utility of the technique is limited, even though there have been instances where between-pass peening proved to be the only solution for a distortion or cracking problem. Before peening is used on a job, engineering approval should be obtained.

Another method for removing shrinkage forces is by thermal stress relieving - controlled heating of the weldment to an elevated temperature, followed by controlled cooling. Sometimes two identical weldments are clamped back to back, welded, and then stress-relieved while being held in this straight condition. The residual stresses that would tend to distort the weldments are thus minimized.

10. Minimize Welding Time

Since complex cycles of heating and cooling take place during welding, and since time is required for heat transmission, the time factor affects distortion. In general, it is desirable to finish the weld quickly, before a large volume of surrounding metal heats up and expands. The welding process used, type and size of electrode, welding current, and speed of travel, thus, affect the degree of shrinkage and distortion of a weldment. The use of mechanized welding equipment reduces welding time and the amount of metal affected by heat and, consequently, distortion. For example, depositing a given-size weld on thick plate with a process operating at 175 amp, 25 volts, and 3 ipm requires 87,500 joules of energy per linear inch of weld (also known as heat input). A weld with approximately the same size produced with a process operating at 310 amp, 35 volts, and 8 ipm requires 81,400 joules per linear inch. The weld made with the higher heat input generally results in a greater amount of distortion. The weld size is, in fact, tied to the heat input. In general, the fillet weld size (in inches) is equal to the square root of the quantity of the heat input (kJ/in) divided by 500. Thus these two welds are most likely not the same size.

Other Techniques for Distortion Control

Water-Cooled Jig

Various techniques have been developed to control distortion on specific weldments. In sheet-metal welding, for example, a water-cooled jig is useful to carry heat away from the welded components. Copper tubes are brazed or soldered to copper holding clamps, and the water is circulated through the tubes during welding. The restraint of the clamps also helps minimize distortion.

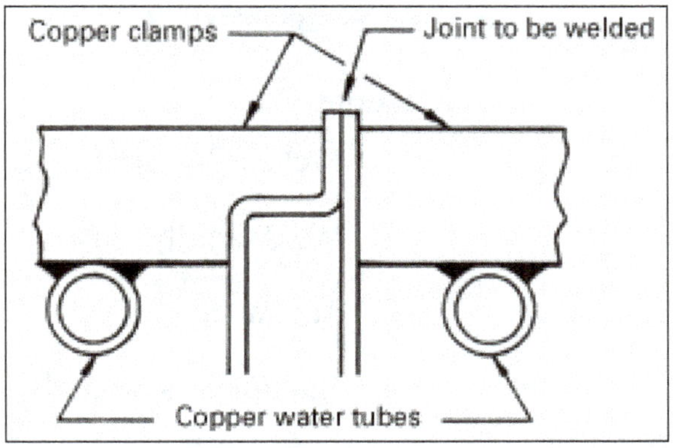

A water-cooled jig for rapid removal of heat when welding sheet meta.

Strongback

The "strongback" is another useful technique for distortion control during butt welding of plates. Clips are welded to the edge of one plate and wedges are driven under the clips to force the edges into alignment and to hold them during welding.

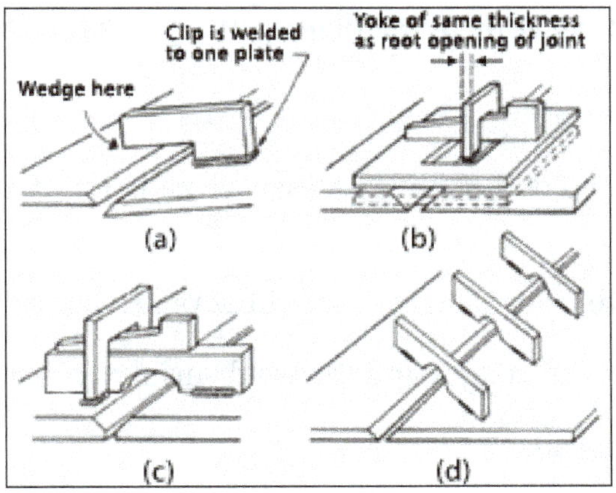

Various strongback arrangements to control distortion during butt-welding.

Thermal Stress Relieving

Except in special situations, stress relief by heating is not used for correcting distortion. There are occasions, however, when stress relief is necessary to prevent further distortion from occurring before the weldment is finished.

How to Avoid Warping of Thin Plates

Acetylene Weld Warping Troubleshooting

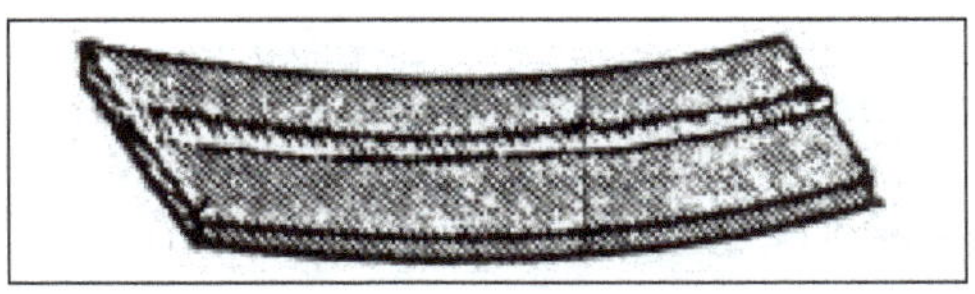

Warping Arc Weld Troubleshooting Suggestions

Why it Occurs:

1. Shrinkage of Material
2. Overheating
3. Faulty Preparation
4. Faulty Clamping of Parts

Solutions:

1. Preheat Material to Relieve Stress
2. Weld rapidly – use back-up weld
3. Too much root gap
4. Clamp parts properly- back-up to cool
5. For multi-layer welds – allow each layer to cool

Step 1. Check for Shrinkage of Deposited Weld Metal

- Select electrode with high welding speed and moderate penetrating properties.

Step 2. Check for Excessive Local Heating at the Joint

- Prevent excessive local heating of the plates adjacent to the weld by welding rapidly.

Step 3. Check for Proper Preparation of Joint

1. In the parts to be welded do not have excessive root opening in the joint between the parts to be welded.
2. Hammer joint edges thinner than the rest of the plates before
3. welding. This elongates the edges and the weld shrinkage causes them to
4. pull back to the original shape.

Step 4. Check the Welding Procedure

1. Use special intermittent or alternating welding sequence and back-step or skip procedure.
2. Preheat material to achieve stress.

Step 5. Check the Clamping of Parts

- Properly clamp parts adjacent to the joint. Use backup fixtures to cool parts rapidly.

How to Avoid Corrosion in Welding

Step 1. Check the Type of Electrode used

1. Bare electrodes produce welds that are less resistant to corrosion than the parent metal.

2. Shield arc electrodes produce welds that are more resistant to corrosion than the parent metal.

3. For the best corrosion resistance, use a filler rod whose composition is similar to that of the base metal.

Step 2. Check to see if the Weld metal Deposited is Proper for the Corrosive Fluid or Atmosphere to be Encountered

- Do not expect more from the weld than you do from the parent metal. On stainless steels, use electrodes that are equal to or better than the parent metal in corrosion resistance.

Step 3. Check on the Metallurgical Effect of the Welding

- When welding 18-8 austenitic stainless steel, be sure the analysis of the steel and welding procedure is correct, so that the welding does not cause carbide precipitations. Carbide precipitation is the rising of carbon to the surface of the weld zone. This condition can be corrected by annealing at 1900 to 2100°F (1038 to 1149°C) after welding. By doing this corrosion in the form of iron oxide, or rust, can be eliminated.

Step 4. Check for Proper Cleaning of the Weld

- Certain materials, such as aluminum, require careful cleaning of all slag after welding to prevent corrosion in service.

How to Prevent Undercutting in your Weld

1. Correct Heat Input

One of the major causes of undercut in the weld is high heat settings while making a run near to the free edges. This may lead to the overheating and melting of near parent metal or previous weld metal. In order to prevent this, one must take great care of weld heat by reducing the current, when approaching to thinner and free edges.

2. Correct Electrode Angle

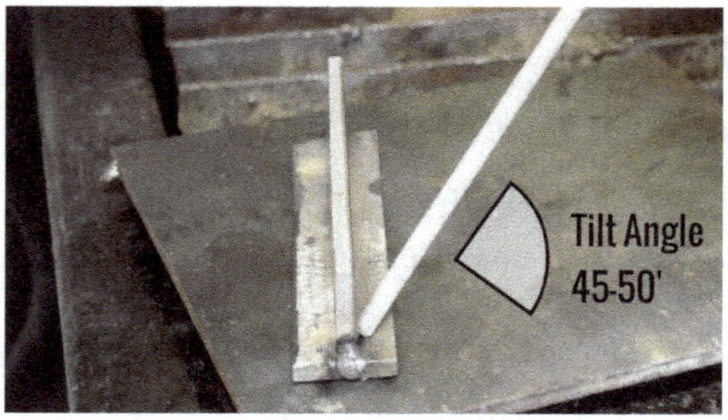

As we know welding angles always play a vital role in producing defects free runs. If one makes a run with an incorrect angle which directs more heat to free edges, here in this case weld is more prone to undercuts. For this one must use correct angles (lead, work, electrode, travel, and slope) in order to direct more heat towards thicker components as compared to the free edges.

3. Moderate Travel Pace

Making a weld with high travel pace is another big cause of weld undercutting. In doing so, some portion of parent metal is drained into the molten metal and never came back due to fast solidification, leaving a depression along the edges. It is recommended to travel a weld run with moderate pace because too slow travel also does not give satisfactory results.

4. Right Selection of Gas Shielding

In case of MAG (metal active gas) process, wrong selection of gas shielding is also one major cause of undercuts in a weld. Welding associates must ensure the correct composition of gas according to the requirement of material type and thickness. Use of Carbon dioxide in conjunction with inert gases provides quality results in welding of carbon steels.

5. Correct welding Technique

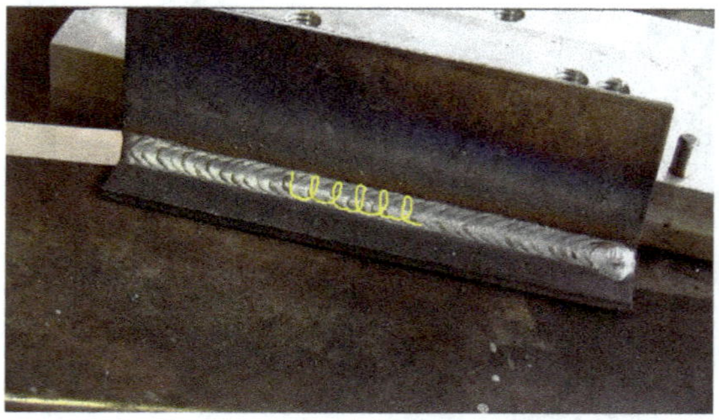

Problem of undercut also arises when a welder endeavors to make a weld run with excessive weaving style. Codes and standards call for the recommended width of weaving according to the electrode size. Size of weaving should not be exceeded to the acceptable limit, otherwise, the weld may prone to undercut grooves. In order to prevent this one must endeavor either to reduce the size of weaving or to produce the weld with multipass technique.

6. Correct Welding Position

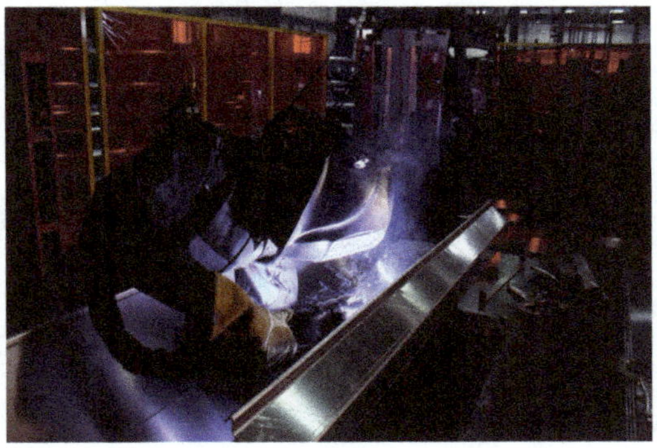

Sometimes making a fillet weld in horizontal-vertical position, also results in weld undercuts. In this case, a depression is formed due to inadequate metal filling along the joint edges. In order to prevent this, try to weld in the flat position to ensure the adequate electrode consumption.

7. Use of Multi-run technique

This one is the best option to adopt, in order to reduce the level of undercuts in welding. Multi-run technique supports all above-mentioned preventions and helps to produce fine welds with improved mechanical properties.

How to Avoid Poor Weld Appearance in Welding

Acetylene Troubleshooting Poor Weld Appearance

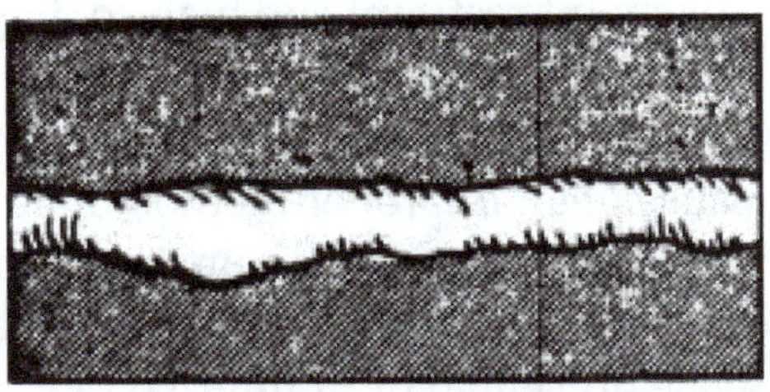

Poor Weld Appearance Arc Weld Troubleshooting

Reasons for Poor Weld Appearance:

1. Uneven Pressure
2. Excessive Stretching
3. Uneven Heating

Solutions:

1. Practice starting, stopping and and finger manipulation on rod
2. Hold rod at proper angle
3. Use slow uniform fanning motion, heat both rod and material

Step 1. Check Welding Technique for Proper Current and Electrode manipulation

1. Ensure the use of the proper welding technique for the electrode used.
2. Do not use excessive welding current.
3. Use a uniform weave or rate of travel at all times.

Step 2. Check Characteristics of type of Electrode used

- Use an electrode designed for the type of weld and base metal and the position in which the weld is to be made.

Step 3. Check Welding Position for which Electrode is Designed

- Do not make fillet welds with down-hand (flat position) electrodes unless the parts are positioned properly.

Step 4. Check for Proper Joint Preparation

- Prepare all joints properly.

Cracking

Cracking may be caused by an improper welding procedure, welder technique, or materials. All types of cracking can be classified as either hot cracking or cold cracking, and these cracks can be oriented transversely or longitudinally to the weld. Transverse cracks are perpendicular to the axis of the weld; longitudinal cracks are parallel to the axis of the weld. Transverse cracks are often the result of longitudinal shrinkage strains acting on excessively hard and brittle weld metal. Longitudinal cracks are often caused by high joint restraint and high cooling rates. Hot cracking is a defect that occurs at higher temperatures and generally happens just after the weld metal starts to solidify. This type of cracking is often caused by excessive sulfur, phosphorous, and lead content in the base metal. It can also occur because of an improper method of breaking the arc or in a root pass when the cross sectional area of the weld bead is small compared to the mass of the base metal. Hot cracking often occurs in deep penetrating welds and it can continue through successive layers if it is not repaired. Hot cracking may be prevented or minimized by:

1. Preheating
2. Using low-hydrogen electrodes
3. Increasing the cross sectional area of the weld bead
4. Changing the contour of the weld bead
5. Using base metal with very low sulfur, phosphorous, and lead content.

Crater cracks are shallow hot cracks caused by improperly breaking the arc.

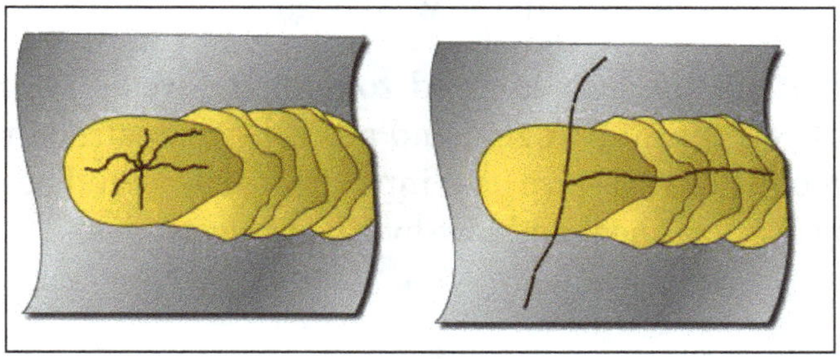

Crater cracks.

Crater cracks may be prevented the same way that craters are, by reversing the travel of the electrode a little way back into the weld from the end of the weld before breaking the arc.

Cold cracking occurs after the weld metal solidification is complete. Cold cracking may occur several days after welding and is generally caused by hydrogen embrittlement, excessive joint restraint, and rapid cooling. Preheating and using low hydrogen electrodes helps reduce this problem. Centerline cracks are cold cracks that often occur in single pass concave fillet welds. A centerline crack is a longitudinal crack that runs down the center of the weld.

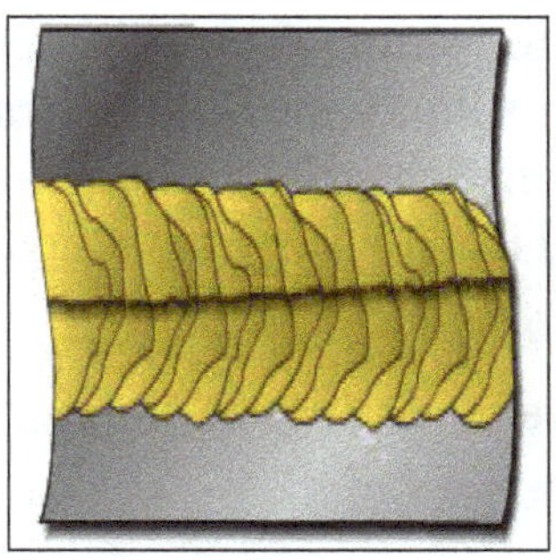

Centerline crack.

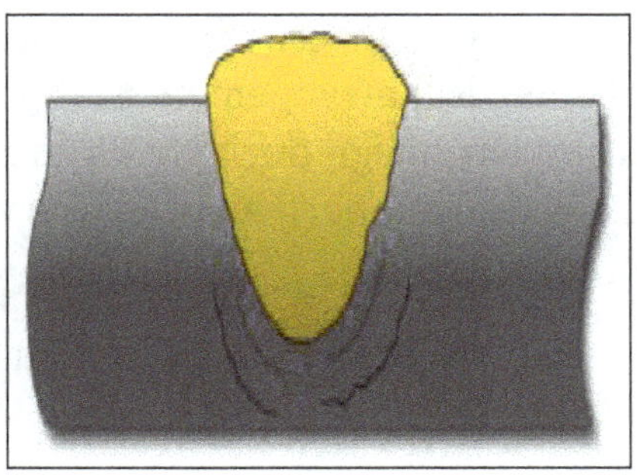

Centerline crack.

This problem may be caused by:

1. Too small a weld bead for the thickness of the base metal
2. Poor fit-up
3. High joint restraint
4. Extension of a crater crack.

The major methods of preventing centerline cracks are:

1. Increasing the bead size
2. Decreasing the gap width
3. Positioning the joint slightly uphill
4. Preventing weld craters.

Base metal and underbead cracks are cold cracks that form in the heat affected zone of the base metal. Underbead cracks occur underneath the weld bead.

Base metal cracks originate in the heat affected zone of the weld. These types of cracking are caused by excessive joint restraint, hydrogen, and a brittle microstructure. A brittle microstructure is caused by rapid cooling or excessive heat input. Underbead and base metal cracking can be reduced or eliminated by using preheat and low-hydrogen electrodes.

Causes of weld cracks:

- Restrained contraction of steel welds during cooling sets up tensile stresses in the joints which can lead to cracks occurring in the weld and HAZ or in both simultaneously. The most common cracks in and around a weld joint are crater cracks, under bead cracks and longitudinal cracks (macro cracks visible to the eye). Cracks along the edge of the weld are sometimes referred to as toe cracks , then there are also hairline cracks along the weld and micro cracks or micro fissuring (not visible to eye and only visible under microscope).
- While these various cracks appear in different parts of the weld and result from different causes, identification and elimination of the basic fault leading to such structural defects will result in crack free welds.
- Weld cracking may also occur from inadequate feeding. The word feeding is used to describe the geometrical arrangement that permits liquid metal to be drawn into and fill any gaps caused by shrinkage during solidification. Generally if the penetration is too deep relative to its width and particularly if the weld is narrower at the top than at its midpoint, shrinkage cracks occurs.
- There are twelve types of cracks which can develop during welding.
    - Arc strike cracking
    - Cold cracking
    - Crater crack
    - Fusion-line cracks
    - Hat crack
    - Hot cracking
    - Under bead crack

- Longitudinal crack
- Reheat cracking
- Root and toe cracks
- Transverse crack

Arc strike cracking:

- Arc strike cracking occurs when the arc is struck but the spot is not welded. This occurs because the spot is heated above the material's upper critical temperature and then essentially quenched. This forms martensite, which is brittle and may lead to higher chances of micro-cracks. Usually the arc is struck in the weld groove so this type of crack does not occur, but if the arc is struck outside of the weld groove then it must be welded over to prevent the cracking. If this is not an option then the arc spot can be post heated, that is, the area is heated with an oxy-acetylene torch, and then allowed to cool slowly.

Crater crack:

- Crater cracks occur when a crater is not filled before the arc is broken. This causes the outer edges of the crater to cool more quickly than the crater, which creates sufficient stresses to form a crack. Longitudinal, transverse and/or multiple radial cracks may form.

Fusion-line cracks:

- The following is an excerpt from a research on Damage Behaviour of Welded Joints with Defects on Fusion Line.
- Two kinds of welded joints damage models, with notch and crack on fusion line, were applied to analyse the influences of the mechanical property heterogeneity and geometric heterogeneity such as intensity matching, base metal rupture strain, base metal strain-hardening exponential and welding line width on the damage distribution and the relationship between damage parameter and rupture parameter. In addition, in the presence of entire damage zone method, the crack extension behaviours of the welded joints were investigated.
- Base metal mechanical property has great effects on the damage distribution of the welded joints. On condition of similar load, the welded joints damage value increases with the decreasing base metal rupture strain and the increasing base metal strain-hardening exponential and intensity matching.
- For under-matched welded joints, the damage concentration zone of the welded joints near the notch will transfer from the welded metal side to the base metal side as the base metal rupture strain decreases and strain-hardening exponential increases. But for over-matched welded joints, the damage concentration zone will transfer from the base metal side to the welded metal side as the base metal rupture strain increases and strain-hardening exponential decreases.
- On the other hand, the varied welded line width has little effect on the damage distribution of the welded joints. For the welded joints with crack on fusion line, when loaded and when the crack mouth opening distance reaches the same level, the maximum damage value in

front of the crack increases as the base metal strain-hardening exponential and intensity matching increase, but decreases as the base metal rupture strain increases.

- The establishment of the failure principle on the welded structure should take mechanical property heterogeneity into serious consideration. The numerical calculations of the crack extension behaviour which is based on the entire damage zone viewpoint reveal that the crack extension path is parallel to the fusion line on condition of plane stress but on condition of plane strain the angle between the crack extension path and the fusion line is approximate 45°.The numerical calculations and analysis indicate that micro-damage parameter and macro -rupture parameter both can describe the stress-strain behaviours, and their relationship intensively relies on mechanical property heterogeneity. The investigation on the micro-damage parameter and macro -rupture parameter is undoubtedly supplemented to the safety evaluation and life-span prediction of the welded joints, which is of great theory significance.

Hat crack

- Hat cracks get their name from the shape of the cross-section of the weld, because the weld flares out at the face of the weld. The crack starts at the fusion line and extends up through the weld. They are usually caused by too much voltage or not enough speed.

Under bead crack

- An undercut crack, also known as a heat-affected zone (HAZ) crack, is a crack that forms a short distance away from the fusion line; it occurs in low alloy and high alloy steel. The exact causes of this type of crack are not completely understood, but it is known that dissolved hydrogen must be present. The other factor that affects this type of crack is internal stresses resulting from unequal contraction between the base metal and the weld metal, restraint of the base metal, stresses from the formation of martensite, and stresses from the precipitation of hydrogen out of the metal.

Longitudinal crack

- Longitudinal cracks run along the length of a weld bead. There are three types: check cracks, root cracks, and full centreline cracks. Check cracks are visible from the surface and extend partially into weld. They are usually caused by high shrinkage stresses, especially on final passes, or by a hot cracking mechanism. Root cracks start at the root and extent part way into the weld. They are the most common type of longitudinal crack because of the small size of the first weld bead. If this type of crack is not addressed then it will usually propagate into subsequent weld passes, which is how full cracks (a crack from the root to the surface) usually form.

Reheat cracking

- Reheat cracking is a type of cracking that occurs in HSLA steels, particularly chromium, molybdenum and vanadium steels, during post heating. The phenomenon has also been observed in austenitic stainless steels. It is caused by the poor creep ductility of the heat affected zone. Any existing defects or notches aggravate crack formation. Things that help prevent reheat cracking include heat treating first with a low temperature soak and then with a rapid heating to high temperatures, grinding or peening the weld toes, and using a two layer welding technique to refine the HAZ grain structure.

Root and toe cracks

- A root crack is the crack formed by the short bead at the root (of edge preparation) beginning of the welding, low current at the beginning and due to improper filler material used for welding. Major reason for happening of these types of cracks is hydrogen embrittlement. These types of defects can be eliminated using high current at the starting and proper filler material. Toe crack occurs due to moisture content present in the welded area, it as a part of the surface crack so can be easily detected. Preheating and proper joint formation is must for eliminating these types of defects.

Transverse crack

- Transverse cracks are perpendicular to the direction of the weld. These are generally the result of longitudinal shrinkage stresses acting on weld metal of low ductility. Crater cracks occur in the crater when the welding arc is terminated prematurely. Crater cracks are normally shallow, hot cracks usually forming single or star cracks. These cracks usually start at a crater pipe and extend longitudinal in the crater. However, they may propagate into longitudinal weld cracks in the rest of the weld.

Cold cracks:

- These cracks may not show up immediately. Sometimes the cracks manifest after 24hrs or more. That is why it is recommended to do the visual and NDT inspection of these welds after 24 hrs. These cracks are called cold cracks or delayed cracks. The cold cracking results in
    - Micro fissures
    - Micro cracks
    - Weld metal cracks

Causes and Cure for Cold Cracking of Weld Metal:

Causes	Cures
High stress and susceptible microstructure	
The reason for high stress and susceptible microstructure are as mentioned below	The solution for high stress and susceptible microstructure are as mentioned below.
• Low amperage current with high weld speed.	• High amperage current with low weld speed.
• Occurs due to low amp current, High TSN (Thermal Severity Number) and low heat input.	• Occurs due to high amp current, low TSN (Thermal Severity Number) with high heat input.
• High Plate thickness with low current.	• High plate thickness with high current.
• No Preheat before welding.	• Preheat the metal required.
• Fast cooling of the weld metal.	• Slow cooling of the weld metal.

<table>
<tr><td>

- Poor Joint design and fit up of the metals to be welded.

Due to all the above reasons high stress occurs and susceptible microstructure occurs on the weld metal and then it finally results in cracking in the weld metal.
</td><td>

- Good joint design and proper fit up of the metals to be welded.

Post heat of the weld metal for stress relief.

These are the solution where the cracks due to stresses and susceptible microstructure can be avoided.
</td></tr>
<tr><td colspan="2">Hydrogen getting Trapped in Weld metal</td></tr>
<tr><td>

Reason for hydrogen getting trapped in weld metal are,

- Dirty, Greasy or contaminated surface and usage of electrodes having higher concentration of hydrogen.
- Due to these reasons hydrogen contamination occurs in weld metal.
- When Hydrogen contamination occurs it results in diffusion of hydrogen in the weld metal.
- And finally it results in trap of Hydrogen in weld metal. When the hydrogen tries to come out of the metal matrix into atmosphere cracks are formed.
</td><td>

Solution for hydrogen getting trapped in weld metal are,

- Usage of clean, wire brush or grinded surface metal for welding and usage of low hydrogen electrode.
- This does not allow the hydrogen to be contaminated in weld metal.
- It avoids diffusion of hydrogen along the grain matrix in the weld metal.
- And finally it results in hydrogen free welding.
</td></tr>
</table>

Liquation Cracks:

- Liquation cracking can occur during fabrication by welding in either the heat affected zone in the parent material, or in previously deposited weld metal during a subsequent run. It results from localized melting at grain or other boundaries, combined with the thermal strains associated with welding.

Hot Cracks:

- Hot cracking is also known as solidification cracking. There are two main reasons for solid cracking,
  - The metal must lack ductility.
  - The tensile stress developed must exceed the corresponding fracture stress.
- Solidification cracks are caused by sulphur (also oxygen and selenium) forming iron sulphide at the grain boundaries which cause reduced grain cohesion.
- When cooling the liquid alloy below its liquid temperature, solid crystals are nucleated and grow until at a certain temperature they join together to form a coherent. At this coherent temperature mechanical strength is acquired and it becomes brittle. When it is further cooled beyond the brittle temperature to the nil – ductility temperature ductility appears.
- The interval between the nil ductility temperature and coherent temperature is known as brittle temperature range. If the brittle temperature range is more weld cracking occurs easily. This can occur immediately after welding. The tensile stresses capable of generating solidification cracking may also rise due to bending.

- The hot cracks occur both on weld metal and as well as in the base metal. The causes and cure for the cracks that occur are mentioned below.

Causes and Cure for Hot Cracking of Base Metal:

Causes	Cure
The reason for hot cracking are, • Presence of sulphur content and sulphur segregation in the metal. • Usage of metal with high carbon and high alloy content. • Metal with high tensile stress which exceeds the corresponding fracture stress. • Absence of preheating. • Excessive weld size.	The solution to avoid hot cracking are, • Usage of low sulphur steel for the purpose of welding. • Usage of metal with low carbon and low alloy content. • Ductile metal with low tensile stress. • Preheating is required. • Improved joint design of the metal.

Causes and Cure for Hot Cracking of Weld Metal:

Cause	Cure
The reason for hot cracking in weld metal are as mentioned below, • Low weld current. • High thermal stress. • Excessive travel speed. • Small electrode. • No or low preheat. • High restraint. • Poor joint preparation. • Unbalanced heat input. • Poor weld profile, fit up and electrode manipulation. • Presence of high sulphur content which forms iron sulphide at grain boundaries. . • Absence of manganese in the weld metal.	The solution for hot cracking in weld metal are as mentioned below, • High weld current. • Low thermal stress. • Decrease travel speed. • Larger electrode. • High preheat. • Eliminate or decrease restraint. • Improved joint design. • Balanced heat input. • Improved weld profile, fit up and electrode manipulation. • Presence of low sulphur content which prevents the formation of iron sulphide at grain boundaries. • Presence of manganese which reacts with sulphur, oxygen and selenium to form globular inclusion prevents the metal from cracks.

How to Prevent Weld Cracks

- Be sure that the welding procedure is such so as to provide good welds of proper fusion.

Avoid trying to weld a high sulphur, high phosphorous steel, try to use steel with the lowest Carbon Equivalent.

- Use a low hydrogen electrode, bake it well to eliminate moisture. If the carbon content of the steel is above 0.45 % use preheat before welding. Do postheat treatmeant and stress relief as per guidance .Use ultra-high argon gas (99.99 %) for protection of weld pool when welding with TIG / MIG.

- Crater cracks may be eliminated by filling the end crater at the end of each weld. Use crater fill function (current downslope) when welding with TIG. Use run off tabs if required.
- Space members evenly so the gap between joints is even. Make sure surfaces to be welded with filler wire are clean, free from dirt, oil, moisture and other contaminants. If using electrode make sure they are well baked in oven at prescribed temperature to eliminate moisture. The structure to be welded should be properly designed and rigid joints eliminated if possible to bring down restraint.
- Maximum restraint is obtained when two rigidly clamped thick plates are joined by a weld of small cross-section, therefore do not use too small welds between heavy plates .Minimum restraint occurs in a weld of relatively large cross section between two close butting thin sheets. Be sure to use welds of sufficient size on all joints.
- Always make sure the weld bead is of sufficient strength to withstand the stresses that might develop from the heat of welding. Shrinkage cracks may form if the weld penetration is too deep relative to its width, and particularly if the weld is narrower at the top than at its midpoint.
- The welding sequence should allow the open ends to move as long as possible. Avoid stringer beads and tack welds, instead use a weaving bead to make full size welds by sections 8 or 10 inch long.

How to Avoid Brittle Welds in Welding

Step 1. Check the Type of Electrode used

- Bare electrodes produce brittle welds. Shielded arc electrodes must be used if ductile welds are required.

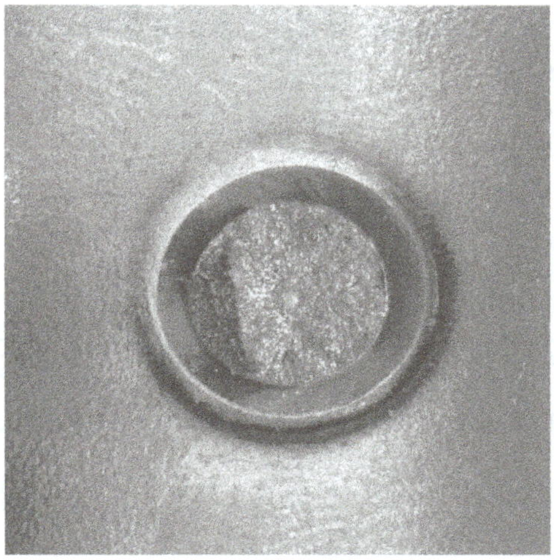

Step 2. Check the Welding Current Setting

- Do not use excessive welding current, as this may cause coarse-grained structure and oxidized deposits.

Step 3. Check for High Carbon or Alloy base Metal which has not been taken into Consideration

1. A single pass weld may be more brittle than a multilayer weld because its micro-structure has not been refined by successive layers of weld metal.
2. Welds may absorb alloy elements from the parent metal and become hard.
3. Do not weld a metal unless the composition and characteristics are known.

How to Avoid Spatter in Welding

Step 1. Check the properties of the electrode used

- Select the proper type of electrode.

Step 2. Check to see if the Welding Current is Excessive for the Type and Diameter of Electrode used

- Use a short arc but do not use excessive welding current.

Step 3. Check for Spalls

1. Paint parts adjacent to welds with whitewash or other protective coating. This prevents spalls from welding to parts, and they can be easily removed.
2. Coated electrodes produce larger spalls than bare electrodes.

CHAPTER 5 Welding Safety

Shielded metal arc welding (SMAW) is a dangerous process and requires strict adherence to proper precautions. Burns, shocks, arc eye along with exposure to UV radiation, gases and particulate matter are some of the common risks associated with this process. The aim of this chapter is to stress on the importance of safety while welding and highlights some of the risks associated with SMAW. It also suggests safety precautions for preventing such injuries.

Safety Practices in Welding

No single task touches on so many facets of safety as welding. Among the issues are fire safety, electric shock, compressed gases, toxic fumes, and personal protection for the eyes, hands, feet, and body. Above all else, caution is crucial when it comes to welding because the risks of personal injury from any mistake are high.

Fire Safety

Fire is always a hazard when it comes to maintenance or repair welding. Prior to welding or torch cutting, you should inspect the work area. Any combustible materials must be moved a minimum of 35 feet from the welding area. Any combustibles that can't be moved must be covered with a fire-resistant shield. Welding sparks and slag fly everywhere, so any holes or cracks in the floor or walls within 35 feet of the welder must be protected.

A fire watch, a person standing nearby with a fire extinguisher, is crucial in the event of a fire. This person should remain in the area for a half-hour after the welding is done. If welding is done after shift workers leave, the guard should be told to pass the area on rounds.

Chemical Safety

Any chemical containers to be welded or cut with a torch must be thoroughly cleaned of flammable materials or greases, tars, acids, or other explosive or toxic materials.

Personal Protection

Welding emits ultraviolet light, so every welder and helper should wear suitable helmets, goggles, or cutting glasses. Lenses are filtered and the amount of the filter is numbered — welders helper's glasses might be a 2.0, brazing goggles might be a 3.0 or 4.0, and the filter for an arc welding helmet might be a 14.

Welding flash, which usually occurs when the welder does not get the hood down in time, is the most common eye injury. Although it is painful when radiation burns the cornea, it's usually not a serious injury. However, watching the welding arc for long periods of time could lead to serious injury.

The face and other parts of the body also need protection from burns, so the hood should cover the face, head, and neck. Shirts should be long sleeved and pockets should be kept buttoned to keep out the sparks. Wool is recommended because of the flame resistance. Man-made fabrics such as polyester, which burns and melts, must be avoided.

A variety of bibs, aprons, capes, and sleeves made of leather or flame-resistant fabric can be used for protection. Footwear should be pull-on high-topped shoes or boots, and flame-resistant gloves should be worn.

Respiratory Protection

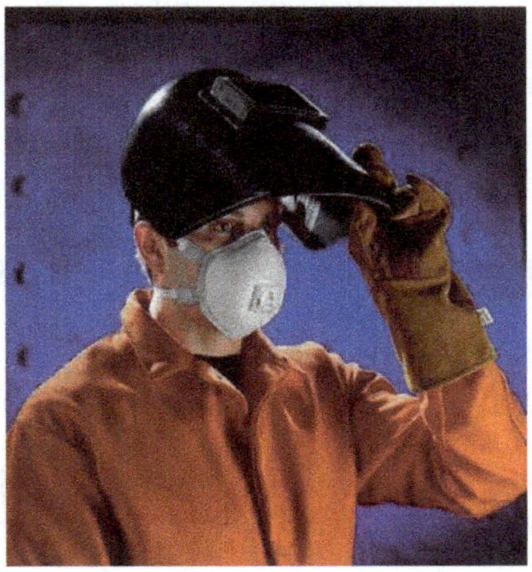

Fumes, a by-product of welding, can be expected from even simple operations and necessitate the use of respiratory protection or ventilation. Simple operations lead to fumes like carbon monoxide, nitrogen oxide, and ozone. But extra precautions must be taken if you are welding metal coated with or containing zinc, cadmium, chromium, copper, fluoride, lead, manganese, or vanadium, because the resulting fumes can cause a condition known as metal-fume fever. Also beware of any metal that has been coated or painted.

Natural ventilation is acceptable if there is space available to each welder of 10,000 cubic feet, the ceiling is more than 16 feet high, the job is not in a confined space, or you are not welding any of the above mentioned materials. Otherwise you will need to provide mechanical ventilation and/or some type of respiratory protection.

Cylinder Safety

When using an oxygen-fuel mixture for welding and cutting, do so with care. Not only is the fuel mixture highly flammable, it is compressed within cylinders.

Acetylene, the most common gas used with oxygen, can be dangerous. So it is important to know the following facts:

- When burned with oxygen, acetylene produces a higher flame temperature (3,300ÞC) than any other gas used commercially;
- The flammable range of acetylene is from 2.5 to 81 percent in air;

- Acetylene is only slightly lighter than air, which means than it neither sinks nor rises rapidly. Instead, it tends to stay in the area where it escaped; and
- In normal atmospheres, acetylene should not be used at a pressure greater than 15 psig, unless connected to a manifold.

As with any compressed gas, the cylinder shall be marked with the chemical or trade name. The cylinders should be stored standing up and be tied, chained, or kept in a rack.

Valve caps or hoods should be over the valve any time the cylinder is not in use. Even empty cylinders have enough pressure to be dangerous if the valve stem is broken.

Roll cylinders on the bottom edge or move with a hand truck, never drag or lift them in a sling.

Oxygen Precautions

When cutting with oxygen, the metal is heated by the burning fuel gas and cut by the reaction of metal and oxygen. While not as flammable as the fuel gases in welding, oxygen should be treated with a lot of respect. Use the same precautions you would with any compressed gas.

Oxygen, as it comes out of a cylinder, has many of the same properties as vapor from liquid oxygen, so keep oil and other petroleum products away. Each oxygen gauge should be marked "Oxygen – Use No Oil."

Never use gauges or regulators that are broken or cannot be read for oxygen, or any compressed gas. Common practice is that the hoses used for oxygen and acetylene are color coded — red for fuel and green for oxygen.

Electrical Safety

Arc welding uses electricity as an energy source, so electric shock is a primary concern of the welder. Accidental contact with live electrical parts is to be avoided. Hazardous situations include damp areas, hot and humid conditions that would cause sweat, and cramped spots where the likelihood of touching conductive parts is high. These problems can be partially solved by insulating the conductive parts near the welder, by the use of dry gloves and clothing in good condition, and by wearing electrically protective footwear.

How to Improve Welding Safety

Becoming a welder requires the utmost in safety at all times. Anyone planning to enter the field or that is already in the field can follow these 12 tips to improve welding safety.

Stay safe by following safe practices and our safety tips. These tips include:

1. Familiarize yourself with the equipment: Every piece of equipment is different despite being similar. You need to read the equipment's operating manual before starting any job to ensure you follow the guidelines set by the manufacturer.

2. Button up: Protecting your skin is essential. This means you want to button up completely. Cuffs, pockets and shirts should be buttoned at all times. Loose material can catch sparks, causing burns.

3. Wear proper clothing: You never want to wear shorts and a t-shirt when welding. Always wear welding jackets, flame-resistant clothing, gloves and the proper eyewear at all times. Even if the material is heavier or hotter, it is worth the added safety it provides.

4. Boots or leather shoes with a high top offer the best foot protection.

5. Proper ventilation is a must at all times: Fumes and smoke pose a health risk and can be toxic. Exhaust hoods and respirators may be required on some jobs.

6. Avoid the light: "Arc flash" can damage the eyes. Always avoid exposure to the flash. Helmets with shields and lens shade work best with OSHA recommendations.

7. Auto-darkening Helmets: Auto-darkening helmets are recommended and will automatically darken the lens using sensors. These helmets are recommended for avoiding "arc flash."

8. Avoid stress injuries: Repetitive tasks, such as snapping the head down to drop the hood of the helmet, can lead to stress injuries. These repetitive tasks should be avoided at all costs.

9. Clean your workspace: Cluttered workspaces lead to injuries and accidents. Your area should contain all the equipment and tools you need.

10. Wire feeders: Proper wire feeders are recommended to add to your personal safety. Boom-mounted wire feeders are optimal because they limit hazards caused by cable clutter.

11. Fixturing: Always optimize fixturing. A gearbox, for example, removes the need for a chain and hoist, which makes the workplace safer as a result.

12. Stick and carrot: Employers should make safety a part of daily life in the workplace. Personal rewards for safety, for example, allow employees to have an incentive to work safely at all times.

How to Select Protective Equipments to Stay Safe

Helmets and Shields

For arc welding, the electric arc is a very powerful source of light, including visible, ultraviolet, and infrared. During all electric welding processes, operators must use safety goggles and a hand shield or helmet equipped with a suitable filter glass to protect against the intense ultraviolet and infrared rays. When others are in the vicinity of the electric welding processes, the area must be screened so the arc cannot be seen either directly or by reflection from glass or metal.

During all oxyacetylene welding and cutting processes, operators must use safety goggles to protect the eyes from heat, glare, and flying fragments of hot metals.

Also be sure to keep MSDS sheets (Material Safety Data Sheets) for all hazerdous materials. Every manufacturer provides MSDS sheets to keep you informed regarding any potential hazards, such as if a respirator is needed when working on a project.

HIGH RISK
Welding in confined areas without ventilation
Welding in awkward postures
Welder exposed to flash
Exposure to gases such as ozone, carbon monoxide and toxic substances
Excessive noise
People other than welder exposed to flash

High Risk Areas to Welding Safety

Welding safety starts with having the right protective gear. This includes:

- Respirator/Welders Mask: There are multiple types of respirators. Buy the one that is made for welders and the type of projects you will be performing. If purchasing a mask with a filter, match the filter to the types of metals and coatings used.
- Keep the area clean and check any gasses for signs of leaks.
- Ventilation: All welding areas should have proper ventilation. Check with OSHA for the up to date standards. Poor ventilation leads to "plume poisoning". If you suspect that be inhaled a toxic plume seek medical help immediately.
- Storage: All flammables should be stored in a flammable liquids locker.
- Eye protection: welding eye protection protects against injuries from debris and from the effects of the ultraviolet light. Different types of helmets are made to protect you when performing different types of welding. These vary by shade number, having a passive or auto-darkening lens (automatically adjusts to welding rays) and comfort/fit.
- Fire protection: Sparks created during the welding process can start fires. For welding Class C extinguishers are often used since these are for electrical fires. Sand and water can also help to extinguish fires.
- Protective Clothing: All skin areas need to be protected to protect against molten metal and sparks.

This includes:

- Long sleeve shirts
- Pants that cover the tops of shoes
- Gloves
- shoes or boots
- Hair is protected with something called a welders beanie
- Leather jackets are also effective for protection from slag and sparks
- Leather aprons provide some protection when sitting down
- Shoe covers called spats protect shoes, something helpful if you are working on a project that produces sparks and slag (molten metal)

Welding Safety Tip: Use pliers when handling metals. If you believe a metal is cool, use the back of the hand and slowing bring it closer to the metal. You'll feel the heat as you get closer if it is too hot to handle.

Prepare for Accidents: Keep a first aid kit on hand that includes bandages and burn spray. Consider an option that exceeds ANSI (American National Standards Institute) and OSHA guidelines such as this first aid kit.

Helmets and Shields

Welding Face Shield and Helmet with Respirator

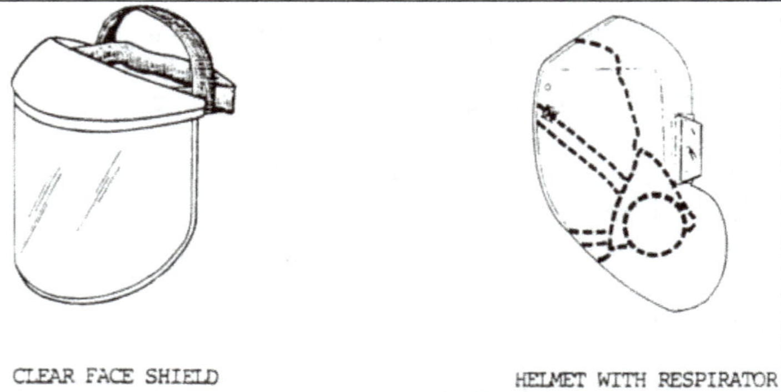

Welding helmets and shields.

Welding Safety Helments and Shields

Welding arcs are intensely brilliant lights.

They contain a proportion of ultraviolet light which may cause eye damage.

For this reason, the arc should never be viewed with the naked eye within a distance of 50.0 ft (15.2 m).

The brilliance and exact spectrum, and therefore the danger of the light, depends on the welding process, the metals in the arc, the arc atmosphere, the length of the arc, and the welding current.

Operators, fitters, and those working nearby need protection against arc radiation.

The intensity of the light from the arc increases with increasing current and arc voltage.

Arc radiation, like all light radiation, decreases with the square of the distance.

Those processes that produce smoke surrounding the arc have a less bright arc since the smoke acts as a filter.

The spectrum of the welding arc is similar to that of the sun. Exposure of the skin and eyes to the arc is the same as exposure to the sun.

Being closest, the welder needs a helmet to protect his eyes and face from harmful light and particles of hot metal.

The welding helmet is generally constructed of a pressed fiber insulating material.

It has an adjustable headband that makes it usable by persons with different head sizes.

To minimize reflection and glare produced by the intense light, the helmet is dull black in color.

It fits over the head and can be swung upward when not welding. The chief advantage of the helmet is that it leaves both hands free, making it possible to hold the work and weld at the same time.

The hand-held shield provides the same protection as the helmet, but is held in position by the handle.

This type of shield is frequently used by an observer or a person who welds for a short period of time.

The protective welding helmet has lens holders used to insert the cover glass and the filter glass or plate.

Standard size for the filter plate is 2 x 4-1/4 in. (50 x 108 mm). In some helmets lens holders open or flip upwards.

Lenses are designed to prevent flash burns and eye damage by absorption of the infrared and ultraviolet rays produced by the arc.

The filter glasses or plates come in various optical densities to filter out various light intensities, depending on the welding process, type of base metal, and the welding current.

The color of the lens, usually green, blue, or brown, is an added protection against the intensity of white light or glare.

Colored lenses make it possible to clearly see the metal and weld. A magnifier lens placed behind the filter glass is sometimes used to provide clear vision.

A cover plate should be placed outside the filter glass to protect it from weld spatter.

The filter glass must be tempered so that is will not break if hit by flying weld spatter.

Filter glasses must be marked showing the manufacturer, the shade number, and the letter "H' indicating it has been treated for impact resistance.

Safety Goggles

Welding Safety Goggles

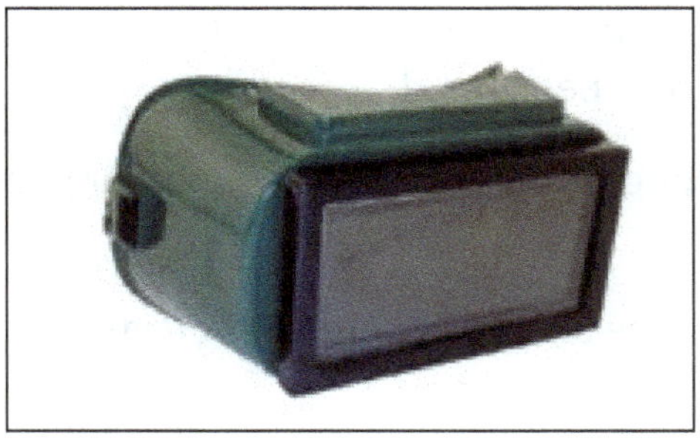

Soft Side Welding Safety Goggles

During all electric welding processes, operators must wear safety goggles.

safety goggles to protect their eyes from weld spatter which occasionally gets inside the helmet.

These clear goggles also protect the eyes from slag particles when chipping and hot sparks when grinding.

Contact lenses should not be worn when welding or working around welders.

Tinted safety glasses with side shields are recommended, especially when welders are chipping or grinding.

Those working around welders should also wear tinted safety glasses with side shields.

Protective Clothing

Welding Protective Clothing

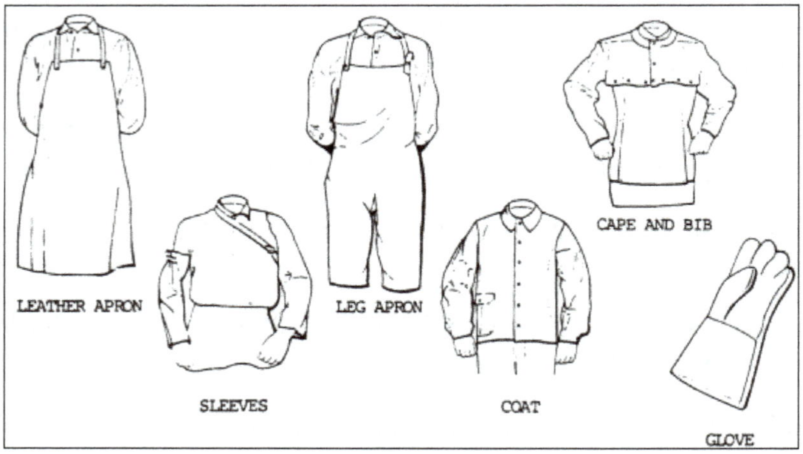

Protective clothing.

Protective Clothing Includes Welding Gloves, Coat, Sleeves, and Leg Protection.

Personnel exposed to the hazards created by welding, cutting, or brazing operations shall be protected by personal protective equipment, Subpart I, Personal Protective Equipment, paragraph 1910.132.

The appropriate protective clothing required for any welding operation will vary with the size, nature, and location of the work to be performed.

Welders should wear work or shop clothes without openings or gaps to prevent arc rays from contacting the skin.

Those working close to arc welding should also wear protective clothing. Clothing should always be kept dry, including gloves.

Woolen clothing should be worn instead of cotton since wool is not easily burned or damaged by weld spatter and helps to protect the welder from changes in temperature.

Cotton clothing, if used, should be chemically treated to reduce its combustibility.

All other clothing, such as jumpers or overalls, should be reasonably free from oil or grease.

Flameproof aprons or jackets made of leather, fire resistant material, or other suitable material should be worn for protection against spatter of molten metal, radiated heat, and sparks.

Capes or shoulder covers made of leather or other suitable materials should be worn during overhead welding or cutting operations.

Leather skull caps may be worn under helmets to prevent head burns.

Sparks may lodge in rolled-up sleeves, pockets of clothing, or cuffs of overalls and trousers.

Therefore, sleeves and collars should be kept buttoned and pockets should be eliminated from the front of overalls and aprons.

Trousers and overalls should not be turned up on the outside. For heavy work, fire-resistant leggings, high boots, or other equivalent means should be used.

In production work welding safety, a sheet metal screen in front of the worker´s legs can provide further protection against sparks and molten metal in cutting operations.

Welding Jacket

Flame Resistant Welding Jacket with Adjustable Cuff and Waist.

Gloves

Welding Safety Gloves

Weld safety gloves for Mig and Stick welding. Stitched with flame-resistant Kevlar.

Flameproof gauntlet gloves, preferably of leather, should be worn to protect the hands and arms from rays of the arc, molten metal spatter, sparks, and hot metal.

Leather gloves should be of sufficient thickness so that they will not shrivel from the heat, burn through, or wear out quickly.

Leather gloves should not be used to pick up hot items, since this causes the leather to become stiff and crack.

Do not allow oil or grease to come in contact with the gloves as this will reduce their flame resistance and cause them to be readily ignited or charred.

Protective Equipment

Protective equipment is at the heart of any welding safety plan. Where there is exposure to sharp or heavy falling objects or a hazard of bumping in confined spaces, hard hats or head protectors must be used.

For welding and cutting overhead or in confined spaces, steel-toed boots and ear protection must also be used.

For welding safety in any area, the operation should be adequately screened to protect nearby workers or passers-by froman the glare of welding.

The screens should be arranged so that no serious restriction of ventilation exists. The screens should be mounted so that they are about 2.0 ft above the floor unless the work is performed at such a low level that the screen must be extended closer to the floor to protect adjacent workers.

The height of the screen is normally 6.0 ft (1.8 m) but may be higher depending upon the situation.

Screen and surrounding areas must be painted with special paints which absorb ultraviolet radiation yet do not create high contrast between the bright and dark areas.

Light pastel colors of a zinc or titanium dioxide base paint are recommended. Black paint should not be used.

Fire Hazards

The fire prevention welding safety and protection plan is the responsibility of welders, cutters, and supervisors.

Approximately six percent of the fires in industrial plants are caused by cutting and welding which has been done primarily with portable equipment or in areas not specifically designated for such work.

The elaboration of basic precautions to be taken for fire prevention during welding or cutting is found in the Standard for Fire Prevention in Use of Cutting and Welding Processes, National Fire Protection Association Standard 51B, 1962.

Some of the basic welding safety precautions for fire prevention in welding or cutting work are given below.

During the welding and cutting operations, sparks and molten spatter are formal which sometimes fly considerable distances.

Sparks have also fallen through cracks, pipe holes, or other small openings in floors and partitions, starting fires in other areas which temporarily may go unnoticed.

For these reasons, welding or cutting should not be done near flammable materials unless every precaution is taken to prevent ignition.

Hot pieces of base metal may come in contact with combustible materials and start fires.

Fires and explosions have also been caused when heat is transmitted through walls of containers to flammable atmospheres or to combustibles within containers.

Anything that is combustible or flammable is susceptible to ignition by cutting and welding.

When welding or cutting parts of vehicles, the oil pan, gasoline tank, and other parts of the vehicle are considered fire hazards and must be removed or effectively shielded from sparks, slag, and molten metal.

Whenever possible, flammable materials attached to or near equipment requiring welding, brazing, or cutting will be removed. If removal is not practical, a suitable shield of heat resistant material should be used to protect the flammable material.

Fire extinguishing equipment, for any type of fire that may be encountered, must be present.

Safety When Welding in Confined Space

Welding safety in confined spaces has its own set of challenges. A confined space is intended to mean a relatively small or restricted space such as a tank, boiler, pressure vessel, or small compartment of a ship or tank.

When welding or cutting is being performed in any confined space, the gas cylinders and welding machines shall be left on the outside. Before operations are started, heavy portable equipment mounted on wheels shall be securely blocked to prevent accidental movement.

Where a welder must enter a confined space through a manhole or other all opening, means will be provided for quickly removing him in case of emergency. When safety belts and life lines are used for this purpose, they will be attached to the welder´s body so that he cannot be jammed in a small exit opening.

An attendant with a preplanned rescue procedure will be stationed outside to observe the welder at all times and be capable of putting rescue operations into effect.

When arc welding is suspended for any substantial period of time, such as during lunch or overnight, all electrodes will be removed from the holders with the holders carefully located so that accidental contact cannot occur.

The welding machines will be disconnected from the power source.

In order to eliminate the possibility of gas escaping through leaks or improperly closed valves when gas welding or cutting, the gas and oxygen supply valves will be closed, the regulators released, the gas and oxygen lines bled, and the valves on the torch shut off when the equipment will not be used for a substantial period of time.

Where practical, the torch and hose will also be removed from the confined space.

After welding operations are completed, the welder will mark the hot metal or provide some other means of warning other workers.

How to Avoid Welding Arc Flash

- The best way is to avoid exposure to the arc flash in the first place by not being in areas where electric arc welding is happening. Obviously, if you're a welder, assisting a welder, or can't avoid being in the area, this is not possible.
- The next best form of protection is placing a barrier between yourself and the welding process. Welding curtains or solid screens prevent the UV radiation from affecting those not directly involved in the job. Again, this is not always possible and certainly not for the person doing the welding.
- Most arc flashes happen to those in the area and not involved in the welding process or to welders in the short period of time between striking the arc and putting the welding screen in place.
- These types of flashes can be prevented through the simple means of wearing safety glasses. Modern safety glasses have lenses made from polycarbonate, which absorbs up to 99% of the damaging UV radiation and also protects your eyes from welding sparks and other debris in the air.
- Additional protection can be gained by using head-mounted rather than hand-held welding screens, and preferably those with flip-up lenses and protective clear polycarbonate screens fixed into the welding screen.
- Auto-darkening welding screens provide the best protection. These screens have special technology that automatically (and almost instantaneously) darkens the lens when UV radiation is detected and then clears when the UV radiation is not present. These types of welding screens also provide much better face and head protection against damaging UV radiation and flying particles generated in the welding process.

How to Avoid Possible Shock Hazards

- Make sure all cables are in good condition with no bare insulation or frayed wires to minimize the extent of live parts.
- Protect cables from vehicular traffic or other hazards so the cables are not damaged, cut, or pinched.
- Make sure that the rod holder is insulated.
- Do not replace the welding electrode with a bare hand, or with a wet welding glove.
- Always keep the hands and the body dry during a welding operation.
- Avoid standing in water, on wet surfaces, or working with wet hands or wearing sweaty garments.
- Do not dip energized (hot) electrode holders in water.
- Avoid direct contact with the live parts of welding equipment and the workpiece.
- Ground the work or metal to be welded to a good electrical ground. Always insulate yourself from the work and ground.
- Wear proper protective equipment, such as rubber boots and rubber pads, if performing arc welding in wet or high humidity conditions. Wear rubber gloves under the welding gloves.
- If the welding operation must be done on steel or other conductive material, use an insulating mat under the operator.
- Place the welding transformer in the close vicinity. In case of need or accident, the transformer can be switched off quickly to cut off the power source.
- Turn off the welding transformer during breaks or not in use. Detach the remaining welding electrode on the electrode holder before leaving the welding area.
- Do not hold or move the welding electrode holder and the welding return cable

simultaneously when moving from one working position to another if the power source to the welding equipment has not been cut.

How to Avoid Fire Hazards

1. Investigate Surroundings before Welding Begins

Fire prevention begins, first and foremost, with the welder. Whether you're inside a shop or out in the field, all welders should investigate the welding area and their surroundings prior to welding. In fact, the welder should clear a distance of 35-feet from the welding area. The appropriate measures should be taken if any welding hazards are identified. These can include removing the fire hazard, relocating the welding project to a different area, or covering the hazard with nonflammable materials. Welding environments, and their surrounding areas, must be kept clean and free of flammable material before and during welding. In most cases, welding operations should only be performed in areas that have been purposely designed to minimize potential fires.

2. Keep Flammable Materials Far from Welding Areas

The sparks and expulsion of molten metal produced by welding and cutting processes are ready sources of ignition that can travel up to 35 feet (10 meters) from their source. Because sparks can travel so far, any combustible material in the immediate area can pose a significant fire hazard. As a result, all welding areas should be free of flammable materials. A flammable material is anything that can be easily set on fire. These include rags, cardboard boxes, paper bags, food, dust, dry leaves, gas cylinders, wood, and cans of paint, solvents, and cleaning products.

In cases where flammable materials cannot be relocated to a safe distance from the welding, then the welding must be moved to a designated safe location. Alternatively, the work area itself can also be enclosed with portable flame-resistant screens that prevent sparks from reaching flammable materials. If neither the flammable material nor the welding can be moved a safe distance from one another, then the flammable material should be covered with tight-fitting, flame-resistant material to prevent ignition. For example, many homes have wood flooring. Wood is a highly-flammable material that must be covered with the appropriate nonflammable material prior to welding.

Utilizing a fire watcher is another alternative for instances where welding cannot be removed from an area with considerable flammable materials. Fire watchers are responsible for monitoring the welding and surrounding areas to ensure there is no sign of fire. Fire watchers continue to monitor the area for at least one half-hour after welding is completed to ensure no possible risk of fire.

3. Cover Holes and Cracks

Many welders don't consider cracks, pipe holes, and other small openings a significant fire hazard. However, flying sparks and molten metal can lodge themselves in small openings in floors and partitions. When this occurs, the material may smolder unseen for seconds, minutes, or even hours before eventually igniting. To prevent sparks from lodging into small openings, identify these areas prior to welding and cover them with the appropriate nonflammable material, move welding to a different location, and/or utilize nonflammable screens.

4. Always Wear the Appropriate PPE

Personal protection equipment, or PPE, refers to garments and equipment that are designed to safeguard the operator's body from physical injury. However, PPE can also prevent fires from occurring in the first place. Welders are directly exposed to flying sparks and molten metal, which can become lodged in the crevices of rolled-up sleeves, collar shirts, pant cuffs, or pockets, resulting in a fire. To prevent such an occurrence, welders should wear long-sleeved shirts and long pants that come over the shoes. Synthetic clothing should be avoided because they are easier to ignite.

5. Always Keep a Fire Extinguisher Nearby

Even if you're very cautious, it's important to be ready for any danger that may occur. As such, all areas where welding is being performed must have a fire extinguisher nearby. Fire extinguishers are very effective in handling the early stages of a fire. When utilized, portable fire extinguishers effectively eliminate 80% of fires. It's important to note that there are different types of fire extinguishers – some are more effective at eliminating certain types of fires than others. For example, Type A fire extinguishers are most effective on combustible solids, such as paper, wood, and clothing. In contrast, Type B fire extinguishers are most effective on combustible liquids, such as oil, grease, and paint thinner. To ensure you get the most from your fire extinguisher, make sure to select the type that suits your particular welding environment. Additionally, all welding personnel must be trained on how to use the fire extinguisher.

CHAPTER 6

Diverse Types of Arc Welding

Science and technology has undergone rapid developments in the past few decades, which has resulted in innovative techniques of arc welding. Some of the variants of arc welding are flux-cored arc welding, gas metal arc welding, gas tungsten arc welding, plasma arc welding, submerged arc welding, etc. which have been discussed in great detail in this chapter.

Flux-cored Arc Welding

Flux cored arc welding is the most productive of the manual welding processes. When comparing MIG welding to flux core arc welding, there is a huge gap in production, in the amount of weld per hour. A MIG welder can typically produce 5 to 8 pounds of weld per hour, versus a FCAW welder packing 25 plus pounds of weld per hour. On top of that flux core welding can weld 1/2″ plates in a single pass with full penetration on both sides. Flux core arc welding for this reason is primarily used in the shipbuilding industry. Ships are made of heavy plate, and have endless amounts of welding that needs to be done. Flux core welding produces high quality welds, fast, and even when in windy conditions.

Process

Flux core welding or tubular electrode welding has evolved from the MIG welding process to improve arc action, metal transfer, weld metal properties, and weld appearance. It is an arc welding process in which the heat for welding is provided by an arc between a continuously fed tubular electrode wire and the workpiece. Shielding is obtained by a flux contained within the tubular electrode wire or by the flux and an externally supplied shielding gas.

The flux-cored welding wire, or electrode, is a hollow tube filled with a mixture of deoxidizers, fluxing agents, metal powders, and ferro-alloys. The closure seam, which appears as a fine line, is the only visible difference between flux-cored wires and solid cold-drawn wire. Flux-cored electrode welding can be done in two ways: carbon dioxide gas can be used with the flux to provide additional shielding, or the flux core alone can provide all the shielding gas and slagging materials. The carbon dioxide gas shield produces a deeply penetrating arc and usually provides better weld than is possible without an external gas shield. Although flux-cored arc welding may be applied semiautomatically, by machine, or automatically, the process is usually applied semiautomatically.

In semiautomatic welding, the wire feeder feeds the electrode wire and the power source maintains the arc length. The welder manipulates the welding gun and adjusts the welding parameters. Flux-cored arc welding is also used in machine welding where, in addition to feeding the wire and maintaining the arc length, the machinery also provides the joint travel. The welding operator continuously monitors the welding and makes adjustments in the welding parameters. Automatic welding is used in high production applications.

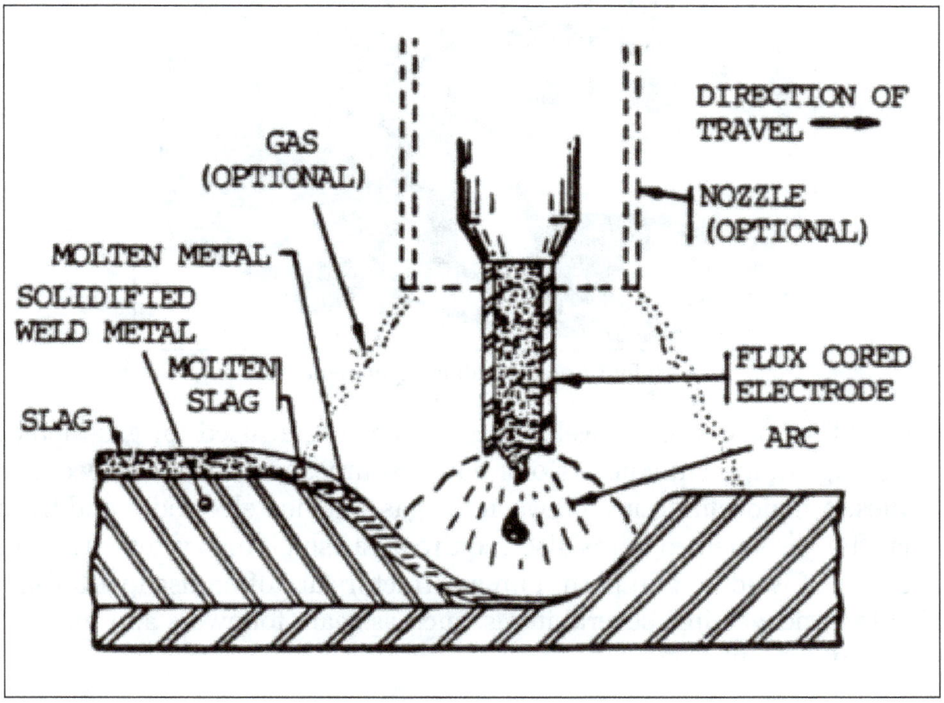

Flux Cored Welding Process Diagram

Welding Tips

- Do not use smooth wire drive rolls, use the knurled drive rolls.
- Change polarity to electrode negative (check with manufacturer, MIG is usually electrode positive).
- Use adequate ventilation.
- 1/2” to 3/4” wire stick out.
- Drag the gun (backhand weld).
- For flat weld, weld at 90 degrees and 10 degrees back. T Joint at 45 degrees. Lap joint at 60 degrees to 70 degrees with one straight weld. For horizontal angle gun upwards at around 10 degrees, turn welding parameters on machine down about 10 to 15%. For vertical weld (can use up or down, vertical down is better for thinner metals, us vertical up for 1/4” and above, also turn parameters down 10 to 15% on machine. For overhead try and maintain a fast travel speed and also reduce welding parameters by 10% to 15% (as compared to flat or horizontal weld).
- Weld side to side to avoid undercut.
- Thoroughly clean off slag after each pass.

Equipment

Flux Cored Welding Equipment

The equipment used for flux core arc welding is similar to that used for gas metal arc welding. The basic arc welding equipment consists of a power source, controls, wire feeder, welding gun, and welding cables. A major difference between the gas shielded electrodes and the self-shielded electrodes is that the gas shielded wires also require a gas shielding system. This may also have an effect on the type of welding gun used. Fume extractors are often used with this process. For machines and automatic welding, several items, such as seam followers and motion devices, are added to the basic equipment.

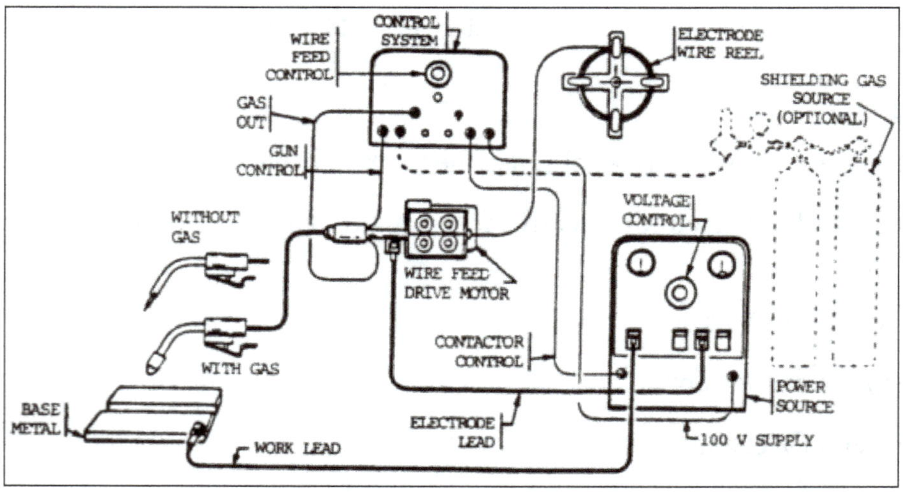

Diagram of Semiautomatic Flux-cored Arc Welding Equipment

Power Source

The power source, or welding machine, provides the electric power of the proper voltage and amperage to maintain a welding arc. Most power sources operate on 230 or 460 volt input power, but machines that operate on 200 or 575 volt input are also available. Power sources may operate on either single phase or three-phase input with a frequency of 50 to 60 hertz. Most power sources used for flux-cored arc welding have a duty cycle of 100 percent, which indicates they can be used to weld continuously. Some machines used for this process have duty cycles of 60 percent, which means that they can be used to weld 6 of every 10 minutes. The power sources generally recommended for flux-cored arc welding are direct current constant voltage type. Both rotating (generator) and static (single or three-phase transformer-rectifiers) are used. The same power sources used with gas metal arc welding are used with flux-cored arc welding. Flux-cored arc welding generally uses higher welding currents than gas metal arc welding, which sometimes requires a larger power source. It is important to use a power source that is capable of producing the maximum current level required for an application.

Direct Current Process

Flux-cored arc welding uses direct current. Direct current can be either reverse or straight polarity. Flux-cored electrode wires are designed to operate on either DCEP or DCEN. The wires designed for use with an external gas shielding system are generally designed for use with DCEP. Some self-shielding flux-cored ties are used with DCEP while others are developed for use with DCEN. Electrode positive current gives better penetration into the weld joint. Electrode negative current gives lighter penetration and is used for welding thinner metal or metals where there is poor fit-up. The weld created by DCEN is wider and shallower than the weld produced by DCEP.

The generator welding machines used for the flux core process can be powered by an electric rotor for shop use, or by an internal combustion engine for field applications. The gasoline or diesel engine-driven welding machines have either liquid or air-cooled engines. Motor-driven generators produce a very stable arc, but are noisier, more expensive, consume more power, and require more maintenance than transformer-rectifier machines.

Wire Feed Motor

A wire feed motor provides power for driving the electrode through the cable and gun to the work. There are several different wire feeding systems available. System selection depends upon the application. Most of the wire feed systems used for flux-cored arc welding are the constant speed type, which are used with constant voltage power sources. With a variable speed wire feeder, a voltage sensing circuit is used to maintain the desired arc length by varying the wire feed speed. Variations in the arc length increase or decrease the wire feed speed. A wire feeder consists of an electrical rotor connected to a gear box containing drive rolls. The gear box and wire feed motor have form feed rolls in the gear box.

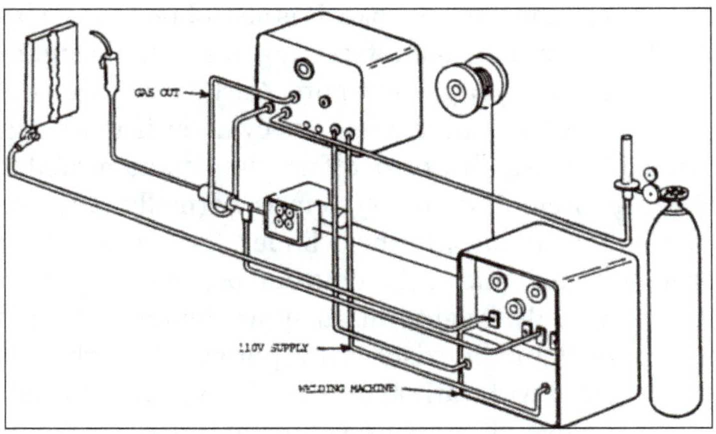

FCAW Wire Feed Assembly

Air and Water Cooled Welding Guns

Both air-cooled and water-cooled guns are used for flux-cored arc welding. Air-cooled flux core guns are cooled primarily by the surrounding air, but a shielding gas, when used, provides additional cooling effects. A water-cooled gun has ducts to permit water to circulate around the contact tube and nozzle. Water-cooled flux core guns permit more efficient cooling of the gun. Water-cooled guns are recommended for use with welding currents greater than 600 amperes, and are preferred for many applications using 500 amperes. Welding guns are rated at the maximum current capacity for continuous operation. Air-cooled guns are preferred for most applications less than 500 amperes, although water-cooled guns may also be used. Air-cooled guns are lighter and easier to manipulate.

Shielding Gases

Shielding gas equipment used for gas shielded flux-cored wires consists of a gas supply hose, a gas regulator, control valves, and supply hose to the welding gun.

The shielding gases are supplied in liquid form when they are in storage tanks with vaporizers, or in a gas form in high pressure cylinders. An exception to this is carbon dioxide. When put in high pressure cylinders, it exists in both liquid and gas forms.

The primary purpose of the shielding gas is to protect the arc and weld puddle from contaminating effects of the atmosphere. The nitrogen and oxygen of the atmosphere, if allowed to come in contact with the molten weld metal, cause porosity and brittleness. In flux-cored arc welding, shielding is accomplished by the decomposition of the electrode core or by a combination of this and surrounding the arc with a shielding gas supplied from an external source. A shielding gas displaces air in the arc area. Welding is accomplished under a blanket of shielding gas. Inert and active gases may both be used for flux-cored arc welding. Active gases such as carbon dioxide, argon-oxygen mixture, and argon-carbon dioxide mixtures are used for almost all applications. Carbon dioxide is the most common. The choice of the proper shielding gas for a specific application is based on the type of metal to be welded, arc characteristics and metal transfer, availability, cost of the gas, mechanical property requirements, and penetration and weld bead shape. The various shielding gases are summarized below.

1. Carbon dioxide: Carbon dioxide is manufactured from fuel gases which are given off by the burning of natural gas, fuel oil, or coke. It is also obtained as a by-product of calcining operation in lime kilns, from the manufacturing of ammonia and from the fermentation of alcohol, which is almost 100 percent pure. Carbon dioxide is made available to the user in either cylinder or bulk containers. The cylinder is more common. With the bulk system, carbon dioxide is usually drawn off as a liquid and heated to the gas state before going to the welding torch. The bulk system is normally only used when supplying a large number of welding stations. In the cylinder, the carbon dioxide is in both a liquid and a vapor form with the liquid carbon dioxide occupying approximately two thirds of the space in the cylinder. By weight, this is approximately 90 percent of the content of the cylinder. Above the liquid, it exists as a vapor gas. As carbon dioxide is drawn from the cylinder, it is replaced with carbon dioxide that vaporizes from the liquid in the cylinder and therefore the overall pressure will be indicated by the pressure gauge. When

the pressure in the cylinder has dropped to 200 psi (1379 kPa), the cylinder should be replaced with a new cylinder. A positive pressure should always be left in the cylinder in order to prevent moisture and other contaminants from backing up into the cylinder. The normal discharge rate of the CO2 cylinder is about 10 to 50 cu ft per hr (4.7 to 24 liters per min). However, a maximum discharge rate of 25 cu ft per hr (12 liters per min is recommended when welding using a single cylinder. As the vapor pressure drops from the cylinder pressure to discharge pressure through the CO2 regulator, it absorbs a great deal of heat. If flow rates are set too high, this absorption of heat can lead to freezing of the regulator and flowmeter which interrupts the shielding gas flow. When flow rate higher than 25 cu ft per hr (12 liters per min) is required, normal practice is to manifold two CO2 cylinders in parallel or to place a heater between the cylinder and gas regulator, pressure regulator, and flowmeter. Excessive flow rates can also result in drawing liquid from the cylinder. Carbon dioxide is the most widely used shielding gas for flux-cored arc welding. Most active gases cannot be used for shielding, but carbon dioxide provides several advantages for use in welding steel. These are deep penetration and low cost. Carbon dioxide promotes a globular transfer. The carbon dioxide shielding gas breaks down into components such as carbon monoxide and oxygen. Because carbon dioxide is an oxidizing gas, deoxidizing elements are added to the core of the electrode wire to remove oxygen.The oxides formed by the deoxidizing elements float to the surface of the weld and become part of the slag covering. Some of the carbon dioxide gas will break down to carbon and oxygen. If the carbon content of the weld pool is below about 0.05 percent, carbon dioxide shielding will tend to increase the carbon content of the weld metal. Carbon, which can reduce the corrosion resistance of some stainless steels, is a problem for critical corrosion application. Extra carbon can also reduce the toughness and ductility of some low alloy steels. If the carbon content in the weld metal is greater than about 0.10 percent, carbon dioxide shielding will tend to reduce the carbon content. This loss of carbon can be attributed to the formation of carbon monoxide, which can be trapped in the weld as porosity deoxidizing elements in the flux core reducing the effects of carbon monoxide formation.Argon-carbon dioxide mixtures.

2. Argon and carbon dioxide are sometimes mixed for use with flux-cored arc welding. A high percentage of argon gas in the mixture tends to promote a higher deposition efficiency due to the creation of less spatter. The most commonly used gas mixture in flux-cored arc welding is a 75 percent argon-25 percent carbon dioxide mixture. The gas mixture produces a fine globular metal transfer that approaches a spray. It also reduces the amount of oxidation that occurs, compared to pure carbon dioxide. The weld deposited in an argon-carbon dioxide shield generally has higher tensile and yield strengths. Argon-carbon dioxide mixtures are often used for out-of-position welding, achieving better arc characteristics. These mixtures are often used on low alloy steels and stainless steels. Electrodes that are designed for use with CO2 may cause an excessive buildup of manganese, silicon, and other deoxidizing elements if they are used with shielding gas mixtures containing a high percentage of argon. This will have an effect on the mechanical properties of the weld.

3. Argon-oxygen mixtures: Argon-oxygen mixtures containing 1 or 2 percent oxygen are used for some applications. Argon-oxygen mixtures tend to promote a spray transfer which

reduces the amount of spatter produced. A major application of these mixtures is the welding of stainless steel where carbon dioxide can cause corrosion problems.

Electrodes

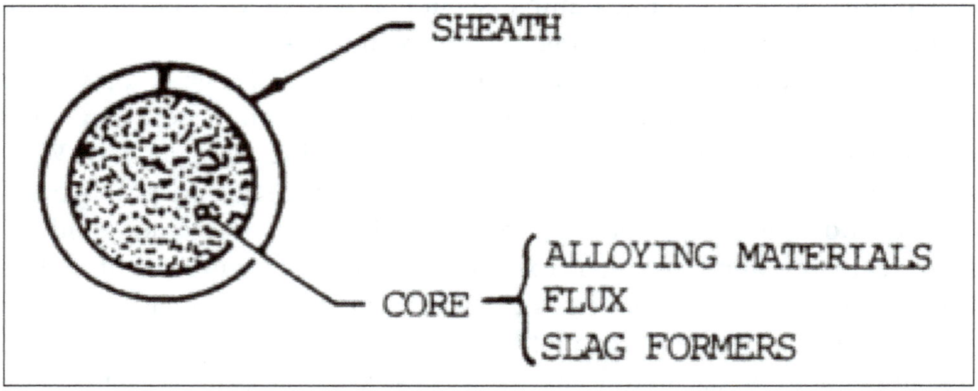

Cross Section of Flux Core Wire

The electrodes used for flux-cored arc welding provide the filler metal to the weld puddle and shielding for the arc. Shielding is required for sane electrode types. The purpose of the shielding gas is to provide protection from the atmosphere to the arc and molten weld puddle. The chemical composition of the electrode wire and flux core, in combination with the shielding gas, will determine the weld metal composition and mechanical properties of the weld. The electrodes for flux-cored arc welding consist of a metal shield surrounding a core of fluxing and/or alloying compounds.

The cores of carbon steel and low alloy electrodes contain primarily fluxing compounds. Some of the low alloy steel electrode cores contain high amounts of alloying compounds with a low flux content. Most low alloy steel electrodes require gas shielding. The sheath comprises approximately 75 to 90 percent of the weight of the electrode. Self-shielded electrodes contain more fluxing compounds than gas shielded electrodes. The compounds contained in the electrode perform basically the same functions as the coating of a covered electrode used in shielded metal arc welding.

These functions are:

1. To form a slag coating that floats on the surface of the weld metal and protects it during solidification.
2. To provide deoxidizers and scavengers which help purify and produce solid weld-metal.
3. To provide arc stabilizers which produce a smooth welding arc and keep spatter to a minimum.
4. To add alloying elements to the weld metal which will increase the strength and improve other properties in the weld metal.
5. To provide shielding gas. Gas shielded wires require an external supply of shielding gas to supplement that produced by the core of the electrode.

Classification System for Tubular Wire Electrodes

The classification system used for tubular wire electrodes used as part of flux core welding was devised by the American Welding Society. Carbon and low alloy steels are classified on the basis of the following items:

1. Mechanical properties of the weld metal.
2. Welding position.
3. Chemical composition of the weld metal.
4. Type of welding current.
5. Whether or not a CO2 shielding gas is used.

An example of a carbon steel electrode classification is E70T-4 where:

1. The "E" indicates an electrode.
2. The second digit or "7" indicates the minimum tensile strength in units of 10,000 psi (69 MPa).
3. The third digit or "0" indicates the welding positions. A "0" indicates flat and horizontal positions and a "1" indicates all positions.
4. The "T" stands for a tubular or flux cored wire classification.
5. The suffix "4" gives the performance and usability capabilities. When a "G" classification is used, no specific performance and usability are required. This classification is intended for electrodes not covered by another classification. Single pass electrodes do not have chemical composition requirements because checking the chemistry of undiluted weld metal does not give the true results of normal single pass weld chemistry.

Welding Cables

The welding cables and connectors are used to connect the power source to the welding gun and to the work. These cables are normally made of copper. The cable consists of hundreds of wires that are enclosed in an insulated casing of natural or synthetic rubber. The cable that connects the power source to the welding gun is called the electrode lead. In semiautomatic welding, this cable is often part of the cable assembly, which also includes the shielding gas hose and the conduit that the electrode wire is fed through. For machine or automatic welding, the electrode lead is normally separate. The cable that connects the work to the power source is called the work lead. The work leads are usually connected to the work by pinchers, clamps, or a bolt.

The size of the welding cables used depends on the output capacity of the flux core welding machine, the duty cycle of the machine, and the distance between the welding machine and the work. Cable sizes range from the smallest AWG No 8 to AWG No 4/0 with amperage ratings of 75 amperes on up.

Table given below shows recommended cable sizes for use with different welding currents and cable lengths. A cable that is too small may become too hot during welding.

Table: Recommended Cable Sizes for Different Welding Currents

Weld Type	Weld Current	Length of Cable Circuit in Feet - Cable Size A.W.G. 60'	100'	150'	200'	300'	400'
Manual (Low Duty Cycle)	100	4	4	4	2	1	1/0
	150	2	2	2	1	2/0	3/0
	200	2	2	1	1/0	3/0	4/0
	250	2	2	1/0	2/0		
	300	1	1	2/0	3/0		
	350	1/0	1/0	3/0	4/0		
	400	1/0	1/0	3/0			
	450	2/0	2/0	4/0			
	500	2/0	2/0	4/0			
Automatic (High Duty Cycle)	400	4/0	4/0				
	800	4/0 (2)	4/0 (2)				
	1200	4/0 (3)	4/0 (3)				

Advantages:

1. FCAW may be an all position process.
2. No shielding gas needed for some wires suitable in windy conditions.
3. Less cleaning of metal is required.
4. Porosity chances are very low.
5. High deposition rate.

Disadvantages:

1. Slag must be removed.
2. More smoke comes out as compared to other processes like GMAW and SAW.
3. Spatter is more.
4. Wire is more expensive.
5. Equipment is expensive and complex.

Application:

1. Used in ship building industry as this saves huge labour cost.
2. Frequently used in construction industries for welding.

FCAW Troubleshooting

When troubleshooting flux core welds, be sure to check the manufacturers directions (found inside the equipment panel) for the following:

- Wire feed speed
- Travel speed
- Contact tip to work distance
- Feeder polarity
- Work angle and travel angle
- Too low of a wire feed feed and current (higher speeds = higher current, lower speeds, lower current: If the speed is too low, you will not get complete coverage, a narrow beed and alot of spatter.

FCAW Weld Created At Low Wire Speed

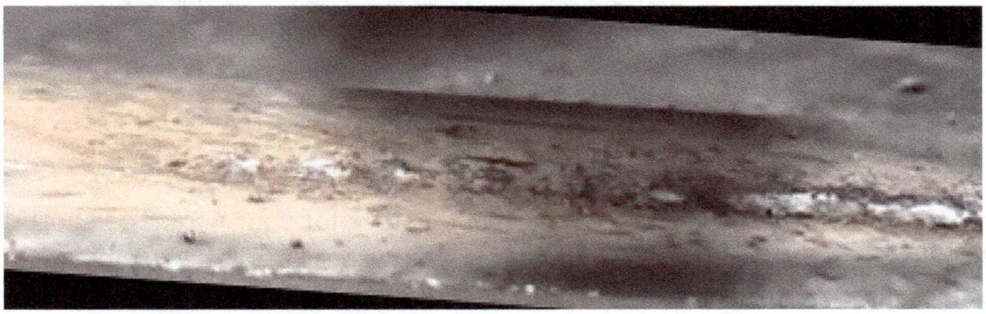

Low wire speed for FCAW weld resulted in hard to remove slag and a lot of spatter.

- If wire speed is too high the wire will keep stubbing. To fix turn voltage up or wire speed down.

FCAW Weld Created at High Wire Speed

- Travel speed too slow: result is a convex wide weld. The slag doesn't cover properly.

FCAW Weld With Low Travel Speed

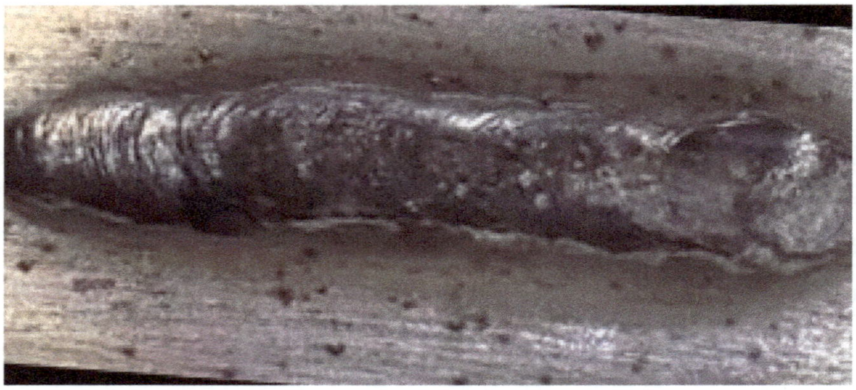

- Travel speed faster than what is recommended: results in a narrow convex weld bead. Compare to too flow travel speed above vs. outrunning puddle below.

FCAW Weld with Fast Travel Speed

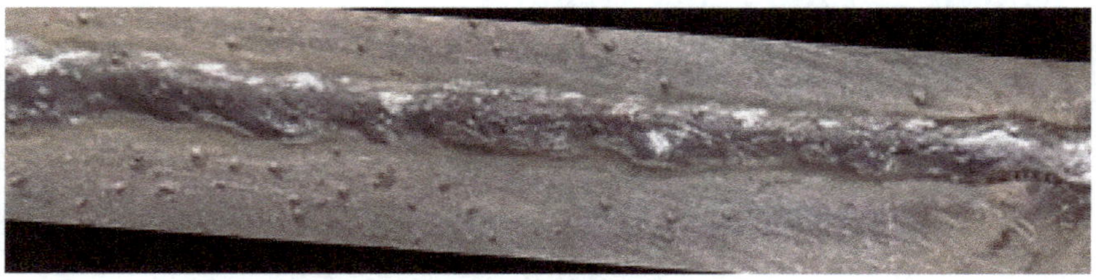

- Contact tip to work distance: Check correct distance for your wire. Too short distance results in inadequate coverage due to the improper preheating of the flux inside the wire. The slag does not cover the entire weld making the slag look dark down the center of the weld. If the distance is too far, there will be some stubbing of the weld. The wire looks like it is hunting for the weld, makes the feeding inconsistent causing ripples in the weld.

FCAW Contact Tip to Work Distance Troubleshooting

Contact tip to work distance is too far (top) and too short (bottom). Check manufacturers directions for correct distance (usually 1/2" to 5/8")

- Polarity: each wire has recommended polarity. Sometimes DC negative is used when DC positive is needed. Causes spatter and a small weld.

FCAW Polarity Troubleshooting

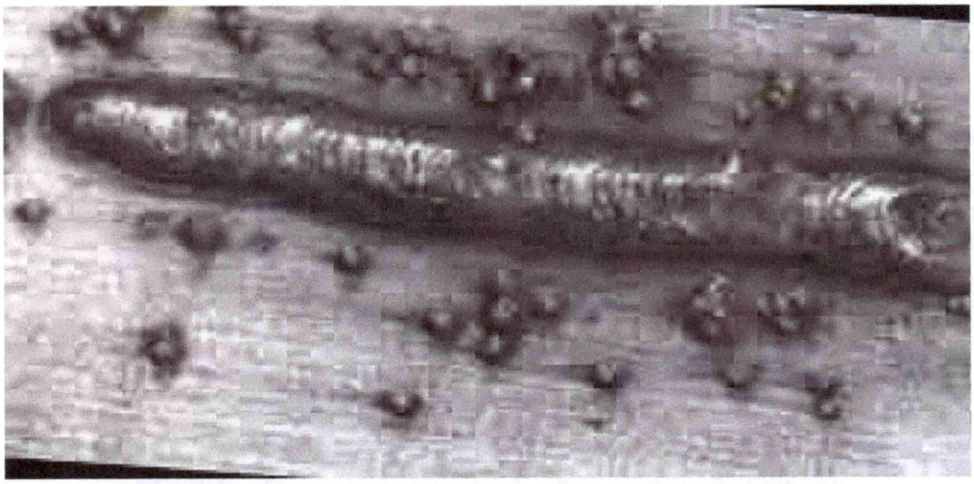

Spatter due to wrong polarity. Make Sure that You are Using the Correct Polarity when Flux Core Welding. Do not use DC Positive if DC Negative is Required. Check Diagram of Machine Setup. Check how Feeder is Connected to Welding Equipment.

FCAW Polarity Feeder Polarity

Make Sure the Feeder is Connected to the Correct Poles. Review Diagram inside of Equipment Panel

- Electrode Angles: For flux core remember that there is slag you drag. Make sure that you drag the electrode to allow the slag to form behind the weld. It is lighter than the molten puddle and it will float to the top. If you push it you chance getting slag inclusions in the weld.
- Check the work and travel angle: If welding on a flat surface, the angle could be 90 degrees. For a lap joint or T joint you want to be 45 degrees to the joint and a 5 to 10 degrees for the drag.

Gas Metal Arc Welding

Gas Metal Arc Welding "Mig" Welding

Gas metal arc welding (GMAW), sometimes referred to by its subtypes metal inert gas (MIG) welding or metal active gas (MAG) welding, is a welding process in which an electric arc forms between a consumable wire electrode and the workpiece metal(s), which heats the workpiece metal(s), causing them to melt and join. Along with the wire electrode, a shielding gas feeds through the welding gun, which shields the process from contaminants in the air.

The process can be semi-automatic or automatic. A constant voltage, direct current power source is most commonly used with GMAW, but constant current systems, as well as alternating current, can be used. There are four primary methods of metal transfer in GMAW, called globular, short-circuiting, spray, and pulsed-spray, each of which has distinct properties and corresponding advantages and limitations.

Originally developed in the 1940s for welding aluminium and other non-ferrous materials, GMAW was soon applied to steels because it provided faster welding time compared to other welding processes. The cost of inert gas limited its use in steels until several years later, when the use of semi-inert gases such as carbon dioxide became common. Further developments during the 1950s and 1960s gave the process more versatility and as a result, it became a highly used industrial process. Today, GMAW is the most common industrial welding process, preferred for its versatility, speed and the relative ease of adapting the process to robotic automation. Unlike welding processes that do not employ a shielding gas, such as shielded metal arc welding, it is rarely used outdoors or in other areas of moving air. A related process, flux cored arc welding, often does not use a shielding gas, but instead employs an electrode wire that is hollow and filled with flux.

Equipment

To perform gas metal arc welding, the basic necessary equipment is a welding gun, a wire feed unit, a welding power supply, a welding electrode wire, and a shielding gas supply.

Welding Gun and Wire Feed Unit

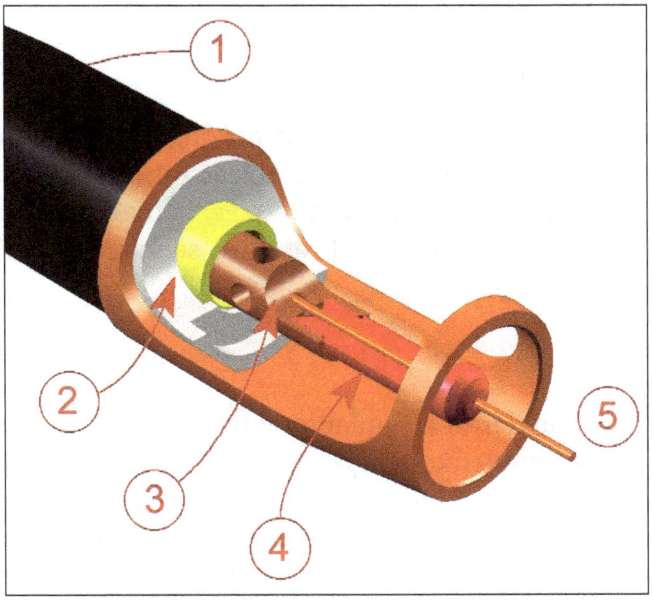

GMAW torch nozzle cutaway image. **(1)** Torch handle, **(2)** Molded phenolic dielectric (shown in white) and threaded metal nut insert (yellow), **(3)** Shielding gas diffuser, **(4)** Contact tip, **(5)** Nozzle output face

GMAW on stainless steel

The typical GMAW welding gun has a number of key parts—a control switch, a contact tip, a power cable, a gas nozzle, an electrode conduit and liner, and a gas hose. The control switch, or trigger, when pressed by the operator, initiates the wire feed, electric power, and the shielding gas flow, causing an electric arc to be struck. The contact tip, normally made of copper and sometimes chemically treated to reduce spatter, is connected to the welding power source through the power cable and transmits the electrical energy to the electrode while directing it to the weld area. It must be firmly secured and properly sized, since it must allow the electrode to pass while maintaining electrical contact. On the way to the contact tip, the wire is protected and guided by the electrode conduit and liner, which help prevent buckling and maintain an uninterrupted wire feed. The gas nozzle directs the shielding gas evenly into the welding zone. Inconsistent flow may not adequately protect the weld area. Larger nozzles provide greater shielding gas flow, which is useful for high current welding operations that develop a larger molten weld pool. A gas hose from the tanks of shielding gas supplies the gas to the nozzle. Sometimes, a water hose is also built into the welding gun, cooling the gun in high heat operations.

The wire feed unit supplies the electrode to the work, driving it through the conduit and on to the contact tip. Most models provide the wire at a constant feed rate, but more advanced machines can vary the feed rate in response to the arc length and voltage. Some wire feeders can reach feed rates as high as 30.5 m/min (1200 in/min), but feed rates for semiautomatic GMAW typically range from 2 to 10 m/min (75 – 400 in/min).

Tool Style

The most common electrode holder is a semiautomatic air-cooled holder. Compressed air circulates through it to maintain moderate temperatures. It is used with lower current levels for welding lap or butt joints. The second most common type of electrode holder is semiautomatic water-cooled, where the only difference is that water takes the place of air. It uses higher current levels for welding T or corner joints. The third typical holder type is a water cooled automatic electrode holder—which is typically used with automated equipment.

Power Supply

Most applications of gas metal arc welding use a constant voltage power supply. As a result, any change in arc length (which is directly related to voltage) results in a large change in heat input and current. A shorter arc length causes a much greater heat input, which makes the wire electrode melt more quickly and thereby restore the original arc length. This helps operators keep the arc length consistent even when manually welding with hand-held welding guns. To achieve a similar effect, sometimes a constant current power source is used in combination with an arc voltage-controlled wire feed unit. In this case, a change in arc length makes the wire feed rate adjust to maintain a relatively constant arc length. In rare circumstances, a constant current power source and a constant wire feed rate unit might be coupled, especially for the welding of metals with high thermal conductivities, such as aluminum. This grants the operator additional control over the heat input into the weld, but requires significant skill to perform successfully.

Alternating current is rarely used with GMAW; instead, direct current is employed and the electrode is generally positively charged. Since the anode tends to have a greater heat concentration, this results in faster melting of the feed wire, which increases weld penetration and welding speed. The polarity can be reversed only when special emissive-coated electrode wires are used, but since these are not popular, a negatively charged electrode is rarely employed.

Electrode

Electrode selection is based primarily on the composition of the metal being welded, the process variation being used, joint design and the material surface conditions. Electrode selection greatly influences the mechanical properties of the weld and is a key factor of weld quality. In general the finished weld metal should have mechanical properties similar to those of the base material with no defects such as discontinuities, entrained contaminants or porosity within the weld. To achieve these goals a wide variety of electrodes exist. All commercially available electrodes contain deoxidizing metals such as silicon, manganese, titanium and aluminum in small percentages to help prevent oxygen porosity. Some contain denitriding metals such as titanium and zirconium to avoid nitrogen porosity. Depending on the process variation and base material being welded the diameters of the electrodes used in GMAW typically range from 0.7 to 2.4 mm (0.028 – 0.095 in) but can be as large as 4 mm (0.16 in). The smallest electrodes, generally up to 1.14 mm (0.045 in) are associated with the short-circuiting metal transfer process, while the most common spray-transfer process mode electrodes are usually at least 0.9 mm (0.035 in).

Shielding Gas

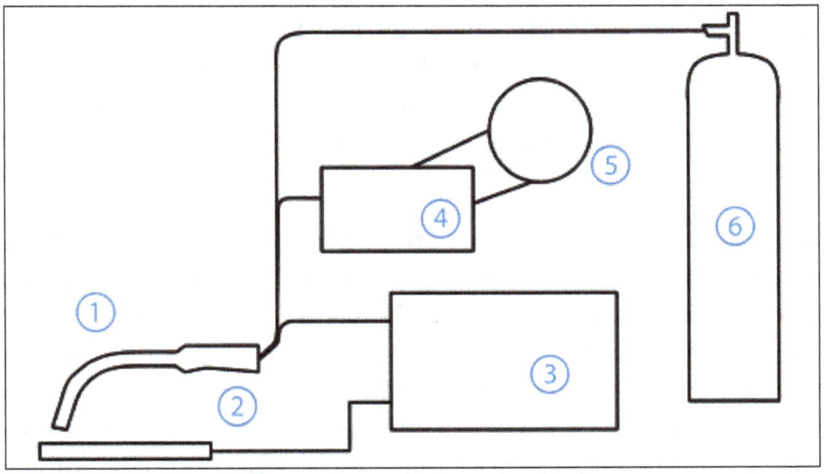

GMAW Circuit diagram. **(1)** Welding torch, **(2)** Workpiece, **(3)** Power source, **(4)** Wire feed unit, **(5)** Electrode source, **(6)** Shielding gas supply.

Shielding gases are necessary for gas metal arc welding to protect the welding area from atmospheric gases such as nitrogen and oxygen, which can cause fusion defects, porosity, and weld metal embrittlement if they come in contact with the electrode, the arc, or the welding metal. This problem is common to all arc welding processes; for example, in the older Shielded-Metal Arc Welding process (SMAW), the electrode is coated with a solid flux which evolves a protective cloud of carbon dioxide when melted by the arc. In GMAW, however, the electrode wire does not have a flux coating, and a separate shielding gas is employed to protect the weld. This eliminates slag, the hard residue from the flux that builds up after welding and must be chipped off to reveal the completed weld.

The choice of a shielding gas depends on several factors, most importantly the type of material being welded and the process variation being used. Pure inert gases such as argon and helium are only used for nonferrous welding; with steel they do not provide adequate weld penetration (argon) or cause an erratic arc and encourage spatter (with helium). Pure carbon dioxide, on the other hand, allows for deep penetration welds but encourages oxide formation, which adversely affects the mechanical properties of the weld. Its low cost makes it an attractive choice, but because of the reactivity of the arc plasma, spatter is unavoidable and welding thin materials is difficult. As a result, argon and carbon dioxide are frequently mixed in a 75%/25% to 90%/10% mixture. Generally, in short circuit GMAW, higher carbon dioxide content increases the weld heat and energy when all other weld parameters (volts, current, electrode type and diameter) are held the same.

As the carbon dioxide content increases over 20%, spray transfer GMAW becomes increasingly problematic, especially with smaller electrode diameters.

Argon is also commonly mixed with other gases, oxygen, helium, hydrogen and nitrogen. The addition of up to 5% oxygen (like the higher concentrations of carbon dioxide mentioned above) can be helpful in welding stainless steel, however, in most applications carbon dioxide is preferred. Increased oxygen makes the shielding gas oxidize the electrode, which can lead to porosity in the deposit if the electrode does not contain sufficient deoxidizers. Excessive oxygen, especially when used in application for which it is not prescribed, can lead to brittleness in the heat affected zone. Argon-helium mixtures are extremely inert, and can be used on nonferrous materials. A helium concentration of 50–75% raises the required voltage and increases the heat in the arc, due to helium's higher ionization temperature. Hydrogen is sometimes added to argon in small concentrations (up to about 5%) for welding nickel and thick stainless steel workpieces. In higher concentrations (up to 25% hydrogen), it may be used for welding conductive materials such as copper. However, it should not be used on steel, aluminum or magnesium because it can cause porosity and hydrogen embrittlement.

Shielding gas mixtures of three or more gases are also available. Mixtures of argon, carbon dioxide and oxygen are marketed for welding steels. Other mixtures add a small amount of helium to argon-oxygen combinations, these mixtures are claimed to allow higher arc voltages and welding speed. Helium also sometimes serves as the base gas, with small amounts of argon and carbon dioxide added. However, because it is less dense than air, helium is less effective at shielding the weld than argon—which is denser than air. It also can lead to arc stability and penetration issues, and increased spatter, due to its much more energetic arc plasma. Helium is also substantially more expensive than other shielding gases. Other specialized and often proprietary gas mixtures claim even greater benefits for specific applications.

The desirable rate of shielding-gas flow depends primarily on weld geometry, speed, current, the type of gas, and the metal transfer mode. Welding flat surfaces requires higher flow than welding grooved materials, since gas disperses more quickly. Faster welding speeds, in general, mean that more gas must be supplied to provide adequate coverage. Additionally, higher current requires greater flow, and generally, more helium is required to provide adequate coverage than if argon is used. Perhaps most importantly, the four primary variations of GMAW have differing shielding gas flow requirements—for the small weld pools of the short circuiting and pulsed spray modes, about 10 L/min (20 ft^3/h) is generally suitable, whereas for globular transfer, around 15 L/min (30 ft^3/h) is preferred. The spray transfer variation normally requires more shielding-gas flow because of its higher heat input and thus larger weld pool. Typical gas-flow amounts are approximately 20–25 L/min (40–50 ft^3/h).

GMAW-based 3-D Printing

GMAW has also been used as a low-cost method to 3-D print metal objects. Various open source 3-D printers have been developed to use GMAW. Such components fabricated from aluminum compete with more traditionally manufactured components on mechanical strength. By forming a bad weld on the first layer, GMAW 3-D printed parts can be removed from the substrate with a hammer.

Operation

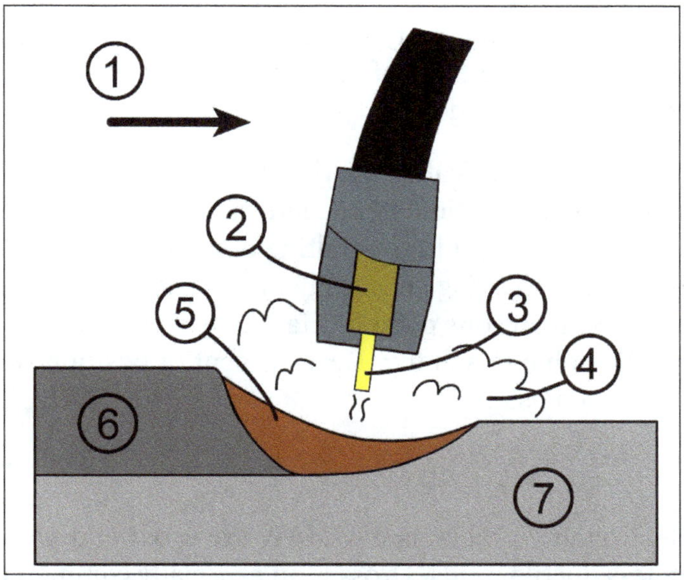

GMAW weld area. **(1)** Direction of travel, **(2)** Contact tube, **(3)** Electrode, **(4)** Shielding gas, **(5)** Molten weld metal, **(6)** Solidified weld metal, **(7)** Workpiece.

For most of its applications gas metal arc welding is a fairly simple welding process to learn requiring no more than a week or two to master basic welding technique. Even when welding is performed by well-trained operators weld quality can fluctuate since it depends on a number of external factors. All GMAW is dangerous, though perhaps less so than some other welding methods, such as shielded metal arc welding.

Technique

GMAW's basic technique is uncomplicated, with most individuals able to achieve reasonable proficiency in a few weeks, assuming proper training and sufficient practice. As much of the process is automated, GMAW relieves the weldor (operator) of the burden of maintaining a precise arc length, as well as feeding filler metal into the weld puddle, coordinated operations that are required in other manual welding processes, such as shielded metal arc. GMAW requires only that the weldor guide the gun with proper position and orientation along the area being welded, as well as periodically clean the gun's gas nozzle to remove spatter buildup. Additional skill includes knowing how to adjust the welder so the voltage, wire feed rate and gas flow rate are correct for the materials being welded and the wire size being employed.

Maintaining a relatively constant contact tip-to-work distance (the *stick-out* distance) is important. Excessive stick-out distance may cause the wire electrode to prematurely melt, causing a sputtering arc, and may also cause the shielding gas to rapidly disperse, degrading the quality of the weld. In contrast, insufficient stick-out may increase the rate at which spatter builds up inside the gun's nozzle and in extreme cases, may cause damage to the gun's contact tip. Stick-out distance varies for different GMAW weld processes and applications.

The orientation of the gun relative to the weldment is also important. It should be held so as to bisect the angle between the workpieces; that is, at 45 degrees for a fillet weld and 90 degrees for welding a flat surface. The travel angle, or lead angle, is the angle of the gun with respect to the direction of travel, and it should generally remain approximately vertical. However, the desirable angle changes somewhat depending on the type of shielding gas used—with pure inert gases, the bottom of the torch is often slightly in front of the upper section, while the opposite is true when the welding atmosphere is carbon dioxide.

Position welding, that is, welding vertical or overhead joints, may require the use of a weaving technique to assure proper weld deposition and penetration. In position welding, gravity tends to cause molten metal to run out of the puddle, resulting in cratering and undercutting, two conditions that produce a weak weld. Weaving constantly moves the fusion zone around so as to limit the amount of metal deposited at any one point. Surface tension then assists in keeping the molten metal in the puddle until it is able to solidify. Development of position welding skill takes some experience, but is usually soon mastered.

Quality

Two of the most prevalent quality problems in GMAW are dross and porosity. If not controlled, they can lead to weaker, less ductile welds. Dross is an especially common problem in aluminium GMAW welds, normally coming from particles of aluminium oxide or aluminum nitride present in the electrode or base materials. Electrodes and workpieces must be brushed with a wire brush or chemically treated to remove oxides on the surface. Any oxygen in contact with the weld pool, whether from the atmosphere or the shielding gas, causes dross as well. As a result, sufficient flow of inert shielding gases is necessary, and welding in moving air should be avoided.

In GMAW the primary cause of porosity is gas entrapment in the weld pool, which occurs when the metal solidifies before the gas escapes. The gas can come from impurities in the shielding gas or on the workpiece, as well as from an excessively long or violent arc. Generally, the amount of gas entrapped is directly related to the cooling rate of the weld pool. Because of its higher thermal conductivity, aluminum welds are especially susceptible to greater cooling rates and thus additional porosity. To reduce it, the workpiece and electrode should be clean, the welding speed diminished and the current set high enough to provide sufficient heat input and stable metal transfer but low enough that the arc remains steady. Preheating can also help reduce the cooling rate in some cases by reducing the temperature gradient between the weld area and the base metal.

Safety

Arc welding in any form can be dangerous if proper precautions are not taken. Since GMAW employs an electric arc, weldors must wear suitable protective clothing, including heavy gloves and protective long sleeve jackets, to minimize exposure to the arc itself, as well as intense heat, sparks and hot metal. The intense ultraviolet radiation of the arc may cause sunburn-like damage to exposed skin, as well a condition known as arc eye, an inflammation of the cornea, or in cases of prolonged exposure, irreversible damage to the eye's retina. Conventional welding helmets contain dark face plates to prevent this exposure. Newer helmet designs feature a liquid crystal-type face plate that self-darkens upon exposure to the arc. Transparent welding curtains, made of a polyvinyl chloride plastic film, are often used to shield nearby workers and bystanders from exposure to the arc.

Welders are often exposed to hazardous gases and airborne particulate matter. GMAW produces smoke containing particles of various types of oxides, and the size of the particles tends to influence the toxicity of the fumes. Smaller particles present greater danger. Concentrations of carbon dioxide and ozone can prove dangerous if ventilation is inadequate. Other precautions include keeping combustible materials away from the workplace, and having a working fire extinguisher nearby.

Metal Transfer Modes

The three transfer modes in GMAW are globular, short-circuiting, and spray. There are a few recognized variations of these three transfer modes including modified short-circuiting and pulsed-spray.

Globular

GMAW with globular metal transfer is considered the least desirable of the three major GMAW variations, because of its tendency to produce high heat, a poor weld surface, and spatter. The method was originally developed as a cost efficient way to weld steel using GMAW, because this variation uses carbon dioxide, a less expensive shielding gas than argon. Adding to its economic advantage was its high deposition rate, allowing welding speeds of up to 110 mm/s (250 in/min). As the weld is made, a ball of molten metal from the electrode tends to build up on the end of the electrode, often in irregular shapes with a larger diameter than the electrode itself. When the droplet finally detaches either by gravity or short circuiting, it falls to the workpiece, leaving an uneven surface and often causing spatter. As a result of the large molten droplet, the process is generally limited to flat and horizontal welding positions, requires thicker workpieces, and results in a larger weld pool.

Short-circuiting

Further developments in welding steel with GMAW led to a variation known as short-circuit transfer (SCT) or short-arc GMAW, in which the current is lower than for the globular method. As a result of the lower current, the heat input for the short-arc variation is considerably reduced, making it possible to weld thinner materials while decreasing the amount of distortion and residual stress in the weld area. As in globular welding, molten droplets form on the tip of the electrode, but instead of dropping to the weld pool, they bridge the gap between the electrode and the weld pool as a result of the lower wire feed rate. This causes a short circuit and extinguishes the arc, but it is quickly reignited after the surface tension of the weld pool pulls the molten metal bead off the electrode tip. This process is repeated about 100 times per second, making the arc appear constant to the human eye. This type of metal transfer provides better weld quality and less spatter than the globular variation, and allows for welding in all positions, albeit with slower deposition of weld material. Setting the weld process parameters (volts, amps and wire feed rate) within a relatively narrow band is critical to maintaining a stable arc: generally between 100 and 200 amperes at 17 to 22 volts for most applications. Also, using short-arc transfer can result in lack of fusion and insufficient penetration when welding thicker materials, due to the lower arc energy and rapidly freezing weld pool. Like the globular variation, it can only be used on ferrous metals.

Cold Metal Transfer

For thin materials, Cold Metal Transfer (CMT) is used by reducing the current when a short circuit is registered, producing many drops per second. CMT can be used for aluminum.

Spray

Spray transfer GMAW was the first metal transfer method used in GMAW, and well-suited to welding aluminium and stainless steel while employing an inert shielding gas. In this GMAW process, the weld electrode metal is rapidly passed along the stable electric arc from the electrode to the workpiece, essentially eliminating spatter and resulting in a high-quality weld finish. As the current and voltage increases beyond the range of short circuit transfer the weld electrode metal transfer transitions from larger globules through small droplets to a vaporized stream at the highest energies. Since this vaporized spray transfer variation of the GMAW weld process requires higher voltage and current than short circuit transfer, and as a result of the higher heat input and larger weld pool area (for a given weld electrode diameter), it is generally used only on workpieces of thicknesses above about 6.4 mm (0.25 in).

Also, because of the large weld pool, it is often limited to flat and horizontal welding positions and sometimes also used for vertical-down welds. It is generally not practical for root pass welds. When a smaller electrode is used in conjunction with lower heat input, its versatility increases. The maximum deposition rate for spray arc GMAW is relatively high—about 600 mm/s (150 in/min).

Pulsed-spray

A variation of the spray transfer mode, pulse-spray is based on the principles of spray transfer but uses a pulsing current to melt the filler wire and allow one small molten droplet to fall with each pulse. The pulses allow the average current to be lower, decreasing the overall heat input and thereby decreasing the size of the weld pool and heat-affected zone while making it possible to weld thin workpieces. The pulse provides a stable arc and no spatter, since no short-circuiting takes place. This also makes the process suitable for nearly all metals, and thicker electrode wire can be used as well. The smaller weld pool gives the variation greater versatility, making it possible to weld in all positions. In comparison with short arc GMAW, this method has a somewhat slower maximum speed (85 mm/s or 200 in/min) and the process also requires that the shielding gas be primarily argon with a low carbon dioxide concentration. Additionally, it requires a special power source capable of providing current pulses with a frequency between 30 and 400 pulses per second. However, the method has gained popularity, since it requires lower heat input and can be used to weld thin workpieces, as well as nonferrous materials.

How to use a MIG Welder

MIG welding is a great way to add professional touches to your DIY projects. MIG welding has many practical applications, from auto work to home repair. Follow the give steps to learn how:

Method 1. Understanding MIG Welding

1. Learn the basics of MIG welding. The process is GMAW (Gas Metal Arc Welding), commonly known as MIG welding (Metal Inert Gas welding). MIG welding was developed during World War II as a fast, portable process to create strong, durable joints. Today it is used in many shop and factory applications as well as by home hobbyists and welding enthusiasts.

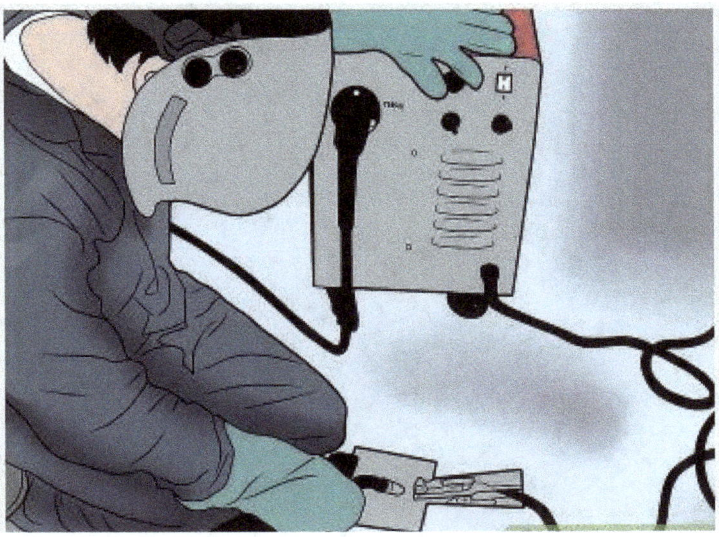

2. Learn how it works. MIG welding uses a machine to feed a wire through a contact tip into a MIG gun. The electrically charged contact tip transfers the welding current to the wire. The arc is established between the wire and the base metal. Often times an inert gas is used, which flows out of the gas nozzle to shield the welding process from the atmosphere. There are several modes of metal transfer:

- Short circuit (thin metals)
- Globular transfer (heavier metals)
- Spray transfer (hottest)

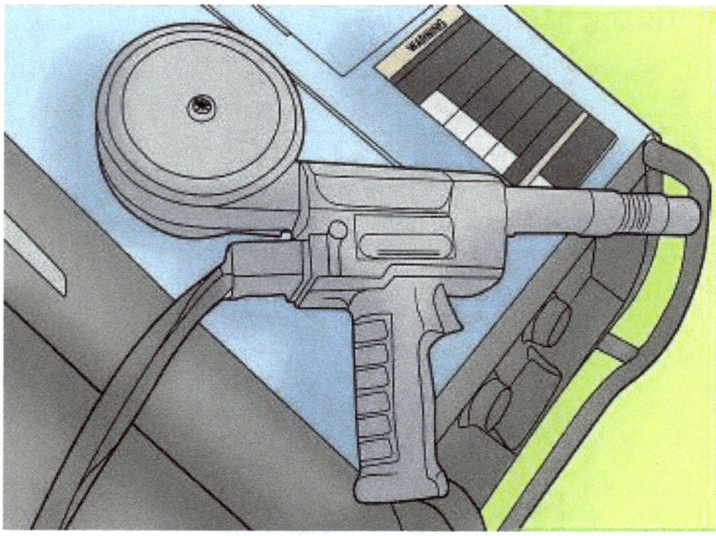

3. Understand the applications. Once you learn how to use a MIG welder, you can make repairs around the home. A MIG welder can be used on stainless steel, mild steel, and aluminium of all thicknesses. Shielding gases will vary depending on base metal and welding wire.

Method 2. Preparing to Weld

1. Assemble your safety gear. You will need a full set of safety equipment to ensure your safety while welding. This includes gloves, masks, and protective clothing.

- Make sure that all of your skin is covered to prevent overexposure from UV rays. You will need a mask with at least a 10 shade or darker. This will help prevent arc eye.
- If you are working in a poorly ventilated area, you will need a vapor mask to minimize the amount of toxic vapors inhaled during the welding process.
- Wear gloves that can protect your skin from molten metal.
- Keep a CO_2 extinguisher and a bucket of sand nearby for emergency fires.

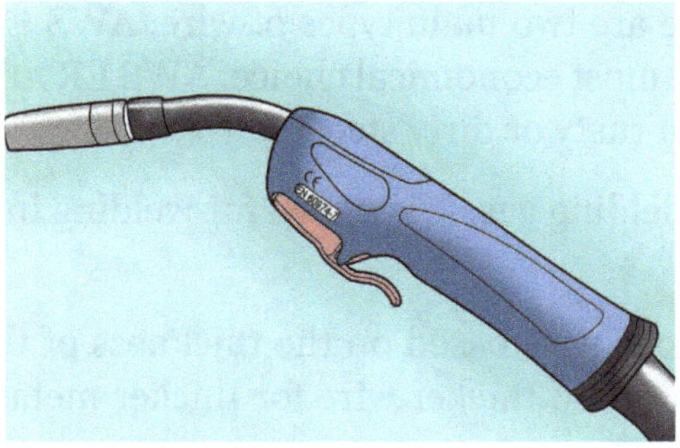

2. Select a comfortable MIG gun. Some are shaped like pistols, while others may look like acetylene torches. The size of the machine will depend on the size of the project.

- A MIG gun may also be water or air cooled. Air-cooled guns are used for 200 amps or less and are easier to manipulate in small areas. The air-cooled gun is the type that home MIG welders typically use.

3. Prepare the area to be welded in. Remove all flammable material and find a good surface to weld on. Although you can put the ground connection right on the piece you are welding, most shops have a large metal workbench that the ground is hooked up to.

- If there are other people present, set up welding curtains around the work area. This will protect them from UV damage.

Method 3. Installing the Wire

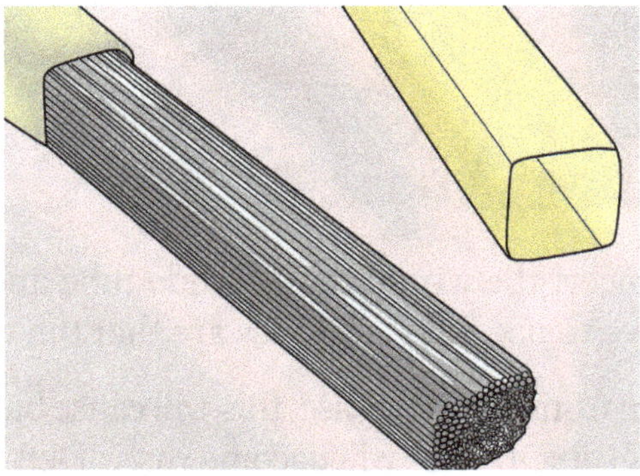

1. Obtain the proper wire. Use the same wire type as the material you are welding. For example, if you are welding stainless steel, use a stainless steel wire.

- For steel welding, there are two main types of wire. AWS ER70S-3 is an all-purpose steel wire. This is usually the most economical choice. AWS ER70S-6 is a high-quality steel wire, designed for welding on rusty or dirty steel.
- E71TGX requires no shielding gas. It is suited for welding in high winds and for painted or rusty materials.
- Vary the diameter of your wire based on the thickness of the metal you are welding. Use thin wire for thin metals, and thicker wire for thicker metals. You may need a larger machine for thicker metals.

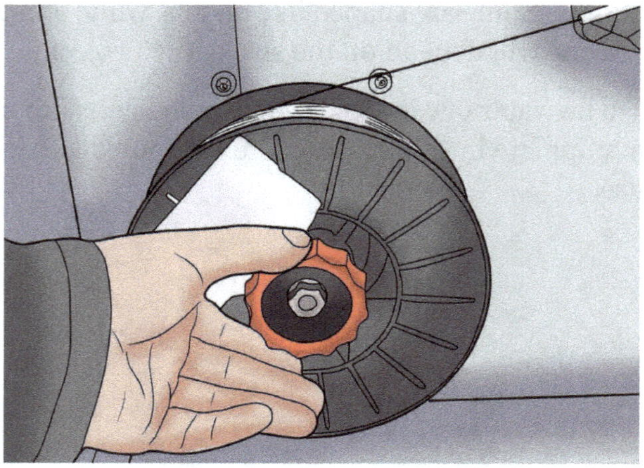

2. Prepare the reel. Tighten the tension on the reel so that the wire doesn't unravel due to its own tension. Make the first 3 inches (7.6 cm) of the wire as straight as possible to avoid tangles or damage to the line feeder. Use a wire cutter to trim the wire accordingly.

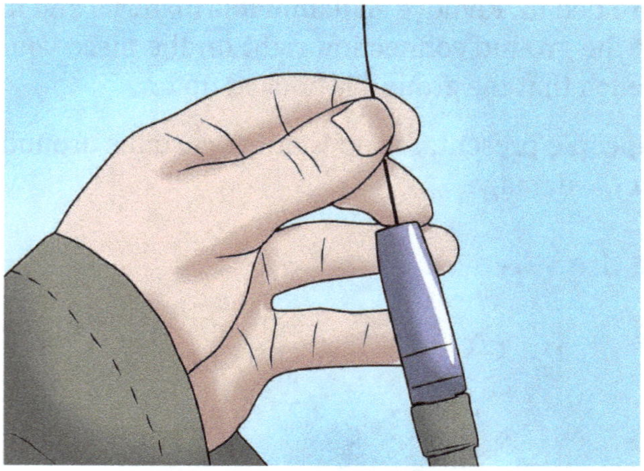

3. Feed the wire to the torch. Insert the wire into the guide tube and feed it over the roller. Insert it in to the wire liner. If you have to use force, chances are that the wire is not aligned properly.

- Make sure the wire is free from rust or grease, this will cause bad welds. Use a dry cloth to clean any dirty wire before inserting it. Wire will become rusty if left in the welder when not in use.
- Once the wire is inserted into the liner, turn the welder on and use the wire feed mechanism to push the wire through the welder.

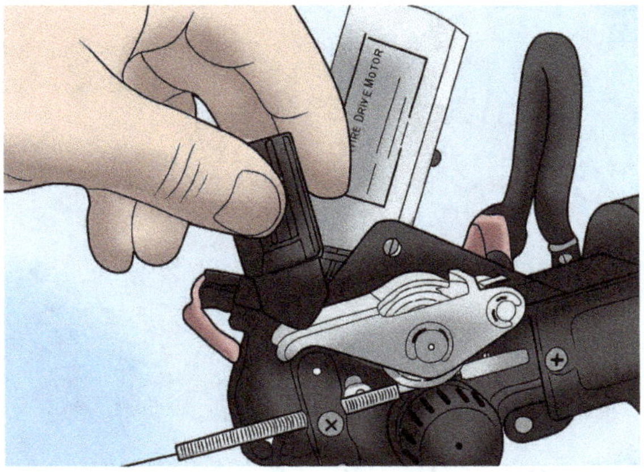

4. Adjust the tension. Once your wire is fed through, you will need to adjust the tensioner. TO much tension will cause the mountings to bend, damaging the welder. Keep the tension at the minimum amount that still allows the line to be fed through.

- Make sure to check the tension on the reel as well as on the line feeder. Both should be as low as possible.

Method 4. Making the Weld

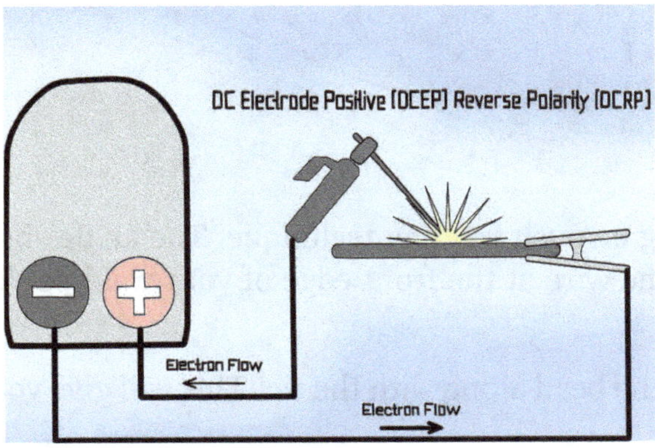

1. Set the welding machine polarity to DCEP. This is reverse polarity.

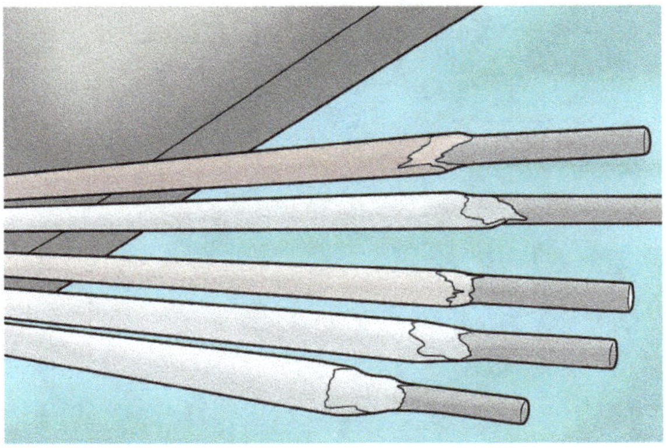

2. Keep a consistent electrode length. As you are welding, keep your electrode extended between 1/4" and 3/8" from the contact tube. This will help make for a clean, regular weld.

3. Use the proper shielding gas. Use carbon dioxide as an economical choice to provide deeper penetration on steel. This will be too hot for thin metals though. Use argon for aluminum welding, and a mix of argon (75%) and carbon dioxide (25%) for thinner steel.

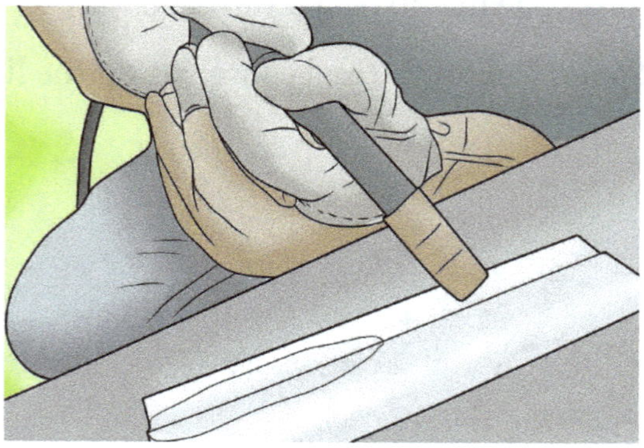

4. Weld a joint using a drag or push welding technique. The angle should not exceed 10 degrees in either technique. Keep the wire at the front edge of your weld pool. This will give you greater control over your weld.

- Drag welding pulls the bead along with the tip. This will give you a deeper penetration and a narrower bead.
- Push welding pushes the bead with the tip. This will give you a wider bead.

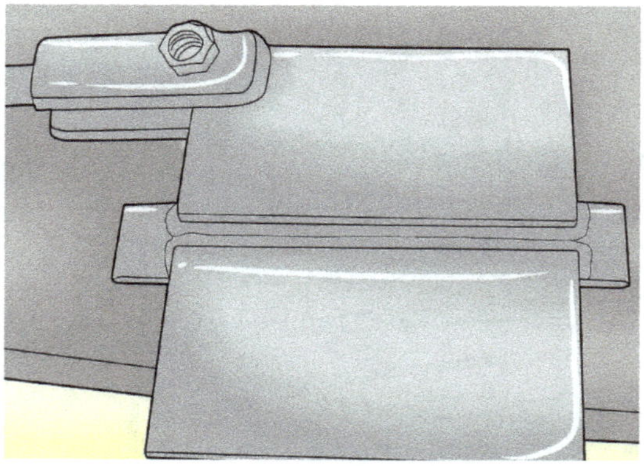

5. Make a flat weld. Use the welder to place material directly into the joint. You can use a back and forth method to fill large gaps. For flat joints, hold the gun at a 90° angle.

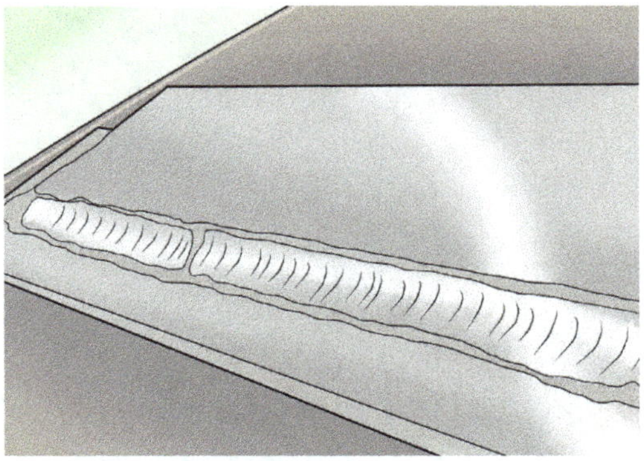

6. Make a horizontal weld. You must lower the gun angle slightly to keep the filler from sagging. Keep the same push or pull angle as normal. Use a back-and-forth weave motion to fill large gaps.

- Keep the amperage the same as a flat weld. You may need to use a slightly smaller diameter wire to keep the weld pool from getting too big.

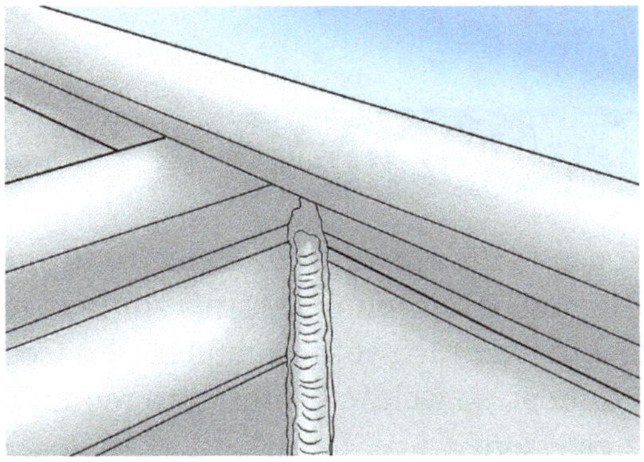

7. Make a vertical weld. For thin materials, start at the top and move the pool down with gravity. This keeps the arc from penetrating the material. For thicker metals, start at the base and work up. This will help increase penetration.

- You may want to lower the amperage by around 10-15% to help fight gravity.

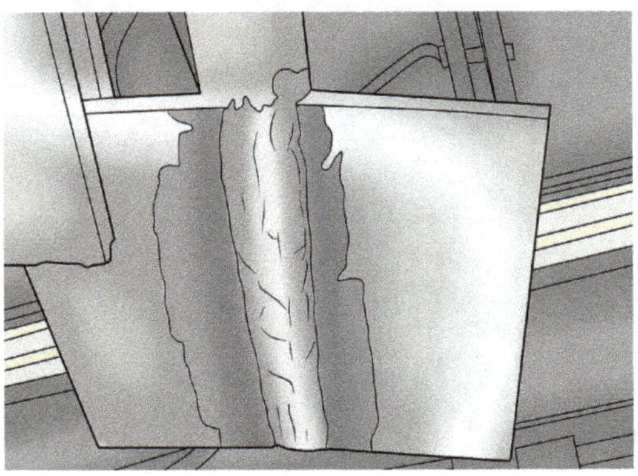

8. Make an overhead weld. Use standard welding techniques, but increase your travel speed. This will help prevent filler from falling out of the joint. You may need to increase your gas flow rate.

- Keep your nozzle clean, as spatter will build up faster when welding overhead.

9. Finish the weld. Once you are done with the weld process, grind off any excess filler. If the weld is defective, grind it down and re-weld the joint.

How to Mig Weld Aluminum

Metal inert gas (MIG) welding uses a consumable wire electrode and shielding gas, which is continuously fed through a welding gun. Aluminum requires some specific changes for welders who are accustomed to welding steel. It's a much softer metal so the feed wire must be larger. Aluminum is also a better conductor of heat, so welding aluminum requires more control over the power supply and the feed rate of the electrode.

Method 1. Choose Equipment and Materials

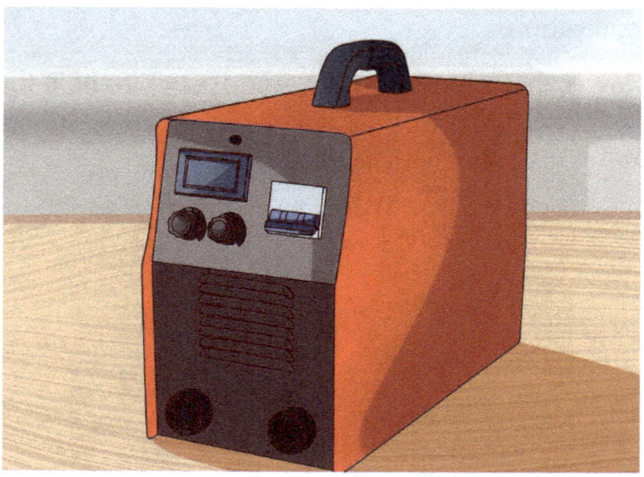

1. Select more powerful welding machines for thicker metal. A 115-volt welder can handle aluminum up to an eighth of an inch thick (3 mm) with adequate preheating, and a 230-volt machine can weld aluminum that's up to a quarter of an inch thick (6 mm). Consider a machine with an output greater than 200 amps if you will be welding aluminum daily.

2. Choose the correct shielding gas. Aluminum requires a shielding gas of pure argon in contrast with steel, which typically uses a blend of argon and carbon dioxide (CO2). This should not require any new hoses, although you may need to replace regulators that were designed specifically for CO2.

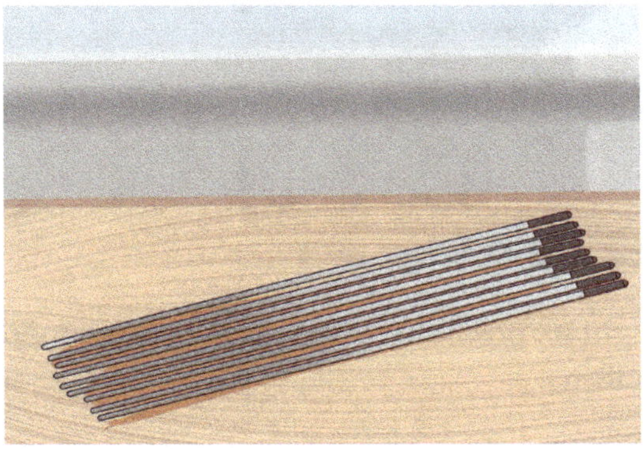

3. Use aluminum electrodes. Electrode thickness is especially critical with aluminum and there is an extremely narrow range to consider. Thinner wire is more difficult to feed, while thicker wire requires greater current to melt. The electrodes for welding aluminum should be .035 of an inch in diameter (less than 1 mm). One of the best choices is 4043 aluminum. A harder alloy like 5356 aluminum is easier to feed, but will require more current.

Method 2. Use Correct Technique

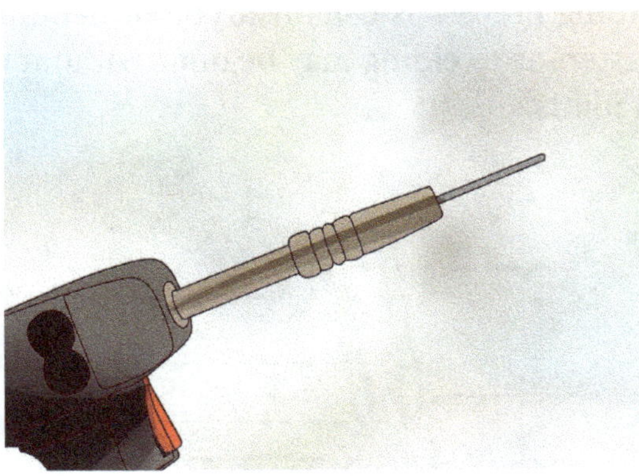

1. Feed the electrodes with an aluminum feeding kit. These kits are commercially available and will will allow you to feed softer aluminum wire with the following features:

- Larger holes on the contact tips. Aluminum expands more than steel as it's heated. This means the contact tips will need larger holes than the ones used for steel wire of the same size. However, the holes should still be small enough to provide good electrical contact.
- U-shaped drive rolls. Aluminum feeders should use drive rolls that won't shave aluminum wire. The inlet and outlet guides for these feeders shouldn't shave the softer aluminum wire. In contrast, steel feeders use V-shaped drive rolls, which are specifically designed to shave the wire.
- Non-metallic liners, which will further reduce the friction on the wire as it goes through the feeder.

2. Keep the gun cable as straight as possible so the wire feeds properly. Softer wire is more prone to kinks due to feeding restrictions.

Gas Tungsten Arc Welding

Gas Tungsten-arc Welding (GTAW) formerly known as TIG (Tungsten Inert Gas) welding, the filler metal is supplied from a filler wire as shown in the figure below. The tungsten electrode is not used during this welding operation, a constant and stable arc gap is maintained at a constant current level. The filler metals are similar to the metals to be welded and flux is not used. The shielding gas used in this welding process is usually argon or helium (or a mixture of these both gases). Welding with gas tungsten-arc welding may be done without using filler metals. for example, in the welding of close-fit joints.

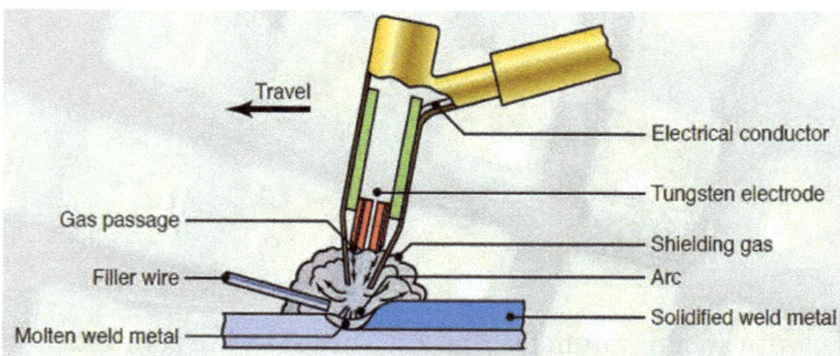

Schematic diagram of Gas Tungsten-arc welding process

Depending on the type of metals to be welded, the power supply is either DC at 200 A or AC at 500 A. In general, AC is preferred for welding metals aluminium and magnesium, because the cleaning action of AC removes oxides and improves weld quality. Thorium or zirconium can be used in the tungsten electrodes to improve their electron emission characteristics. The power supply ranges from 8 to 20 kW. Contamination of the tungsten electrode by the molten metal can be a major problem, particularly in critical applications, because it can cause discontinuities in the weld. Contact of the electrode with the molten-metal pool should be avoided.

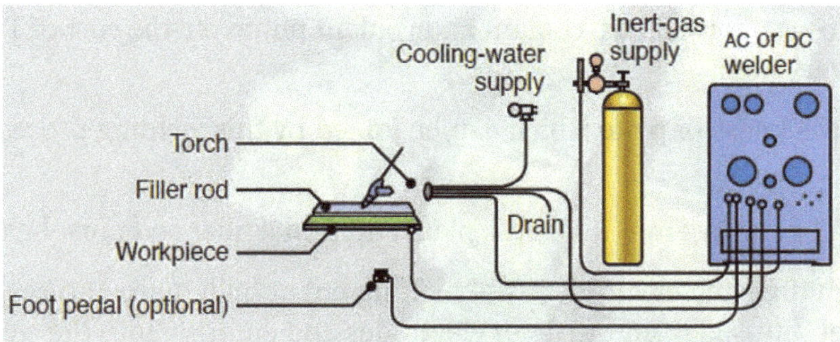

Equipments used for gas tungsten-arc welding operations

The gas tungsten-arc welding process is used for a wide variety of applications and metals, particularly aluminium, magnesium, titanium and the refractory metals. It is highly suitable for thin metals. The cost of the inert gas makes this process more expensive than Shielded Metal-arc Welding but provides welds of very high quality and surface finish. The equipment used for gas tungsten-arc welding process is portable.

Applications of Gas Tungsten-arc Welding

- Originally developed for welding Aluminium and Magnesium.
- The other metals are Stainless steel, High carbon steel, Copper, Monel(Ni + Cu+ Fe + Mg)), Inconel (Cu + Cr + Fe), Brass, Bronze, Silver, Molybdenum etc.
- This process is used for joining various combinations of dissimilar metals like brazing and braze welding.
- Pipe work required for high pressure steam lines, chemical and petroleum industries.
- Welding of air craft frame, jet engine casing, rocket motor casing.
- Accuracy welding of parts in atomic energy.
- Expansions bellows, transistors cases, instrument diagrams etc.

Advantages of Gas Tungsten-arc Welding

- Gas Tungsten-arc welds are stronger and more ductile.
- No danger corrosion due to no flux is used.
- No post weld cleaning because of no slag.
- Wide variety of joints can be made because no flux is used.

- There is very little or no smoke, fumes or sparks at all. This helps in making a neat and sounder weld.
- As the shielding gas is transparent, operator can clearly observe the weld.
- Fusion welds can be made in merely all commercial metals.

Limitations of Gas Tungsten-arc Welding

- Because of usage of inert gas, coolant and coolant pump etc the cost of Tungsten Inert Gas welding is very high.
- Maximum thickness of plate which can be joined by this welding process directly is up to 5mm.
- For welding of above 5 mm thickness plate additional filler rod must be used.
- Even though tungsten electrode is not melting but at high temperature the atoms of tungsten may get diffused from the tip of electrodes and entering into the weld pool which will increase the brittleness of weld bead.

How to TIG Weld

In Tungsten Inert Gas (TIG) welding, a tungsten electrode is used to heat the metal, while Argon gas protects the weld puddle from airborne contaminants. TIG welding can be used to produce high-quality, clean welds on most materials, including steel, stainless steel, chromoly, aluminum, nickel alloys, magnesium, copper, brass, bronze, and gold. Follow the steps below to get your TIG welder up and running and start welding masterpieces today.

Method 1. Setting Up the TIG Machine

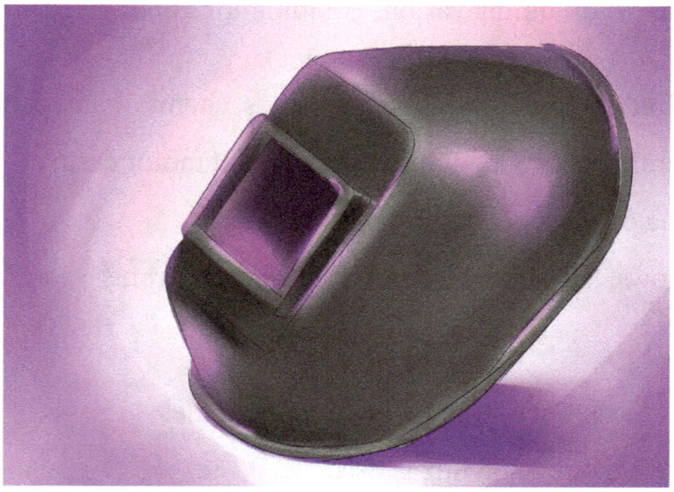

1. Put on safety gear. Before operating any welding machine, make sure to put on protective eyewear, thick, fire-resistant clothing, and a welding helmet with an eye shield.

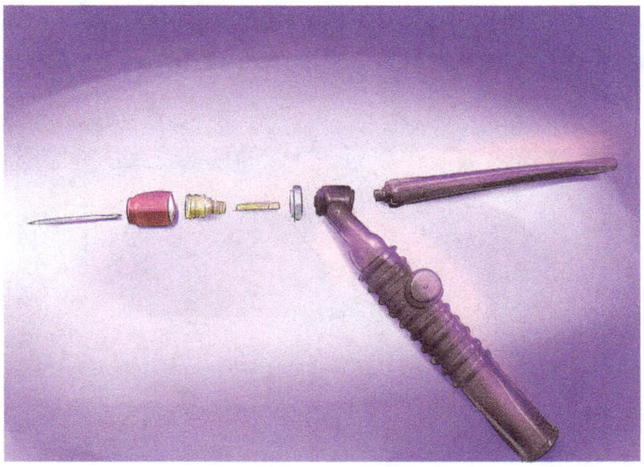

2. Connect the TIG torch. All TIG torches have a ceramic nozzle for directing argon, a copper sleeve for holding an electrode, and some way of cooling themselves. Use the adapter from your accessory package to plug the torch into the front of your machine.

3. Plug your foot pedal into the machine. The foot pedal is used to control the heat at which you are welding.

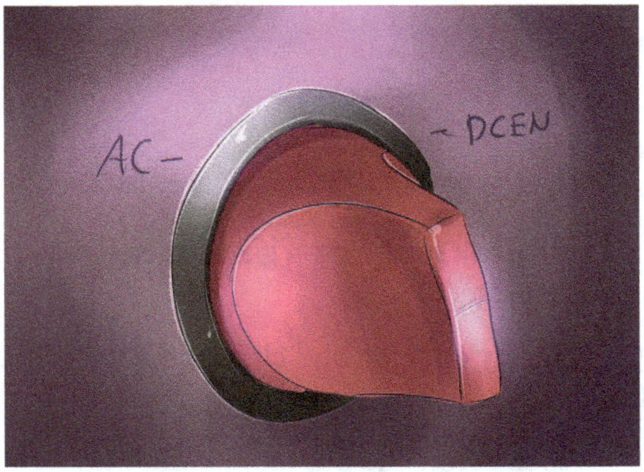

4. Select the polarity. You will pick different settings based on the type of metal that you are welding. If you are using aluminum, place the welder on the alternating current (AC) setting. If you are using steel or other metals, place the welder on the DC Electrode Negative (DCEN) setting.

- If your welder has a high frequency setting, it will also need adjusting. For aluminum, the switch will need to be on continuous high frequency. For steel, it should be on high frequency start.

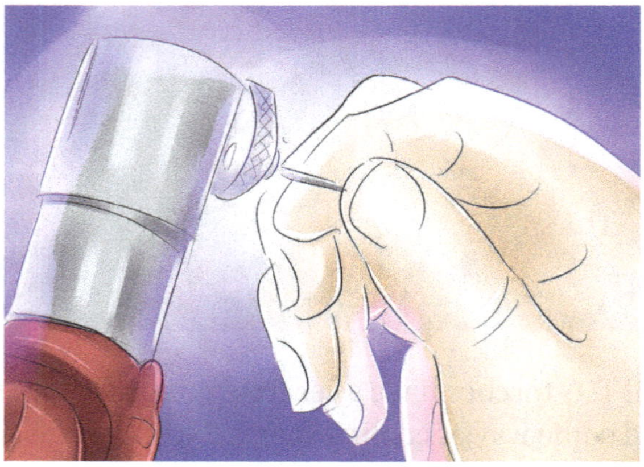

5. Grind the tungsten. The thickness of the metal to be welded and the welding current used determine the size of the tungsten rod. Make sure to grind in a radial direction around the circumference of the tungsten, not straight toward the ends.

- Use the face of a fine stone to grind the tungsten electrode. Grind so that the electrode is pointing in the same direction as the rotation of the stone as a safety precaution.
- Grind the tungsten to a balled tip for AC welding and a pointed tip for DC welding.
- To make a butt weld or open corner weld ground the tungsten to a five to six millimeter stick.

6. Set up the gas flow. You want to use a pure Argon gas or mixed Argon gas such as an Argon-Helium mixture. Remove the plastic protective cap.

- Shift the valve body by quickly opening and closing the valve to clean any debris out of the threaded valve body.
- Screw the regulator on, then screw the nut tight while simultaneously twisting the regulator until it is seated in the valve.
- Tighten the regulator using a spanner, making sure that the pressure knob is backed off counter-clockwise.

- Put on the gas hose and flowmeter, then turn on the cylinder valve. Make sure to turn on the cylinder valve gently and in small increments. Usually a quarter-turn revolution is enough.
- Finally, check for any leaks by listening for a wheezing sound or using an aerosol leak detector spray.
- Set the gas flow rate by adjusting the cylinder regulator. Though the rate may vary depending on your project, usually the rate stays between four and 12 liters (3.2 US gal) per minute.

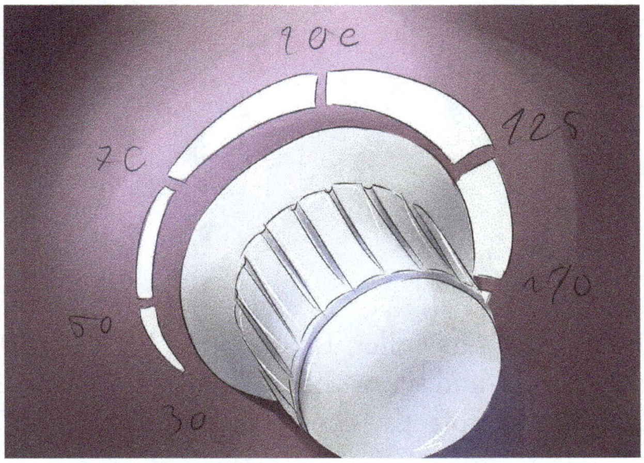

7. Set the amperage. The amperage allows you to regulate the control you have over the welding process.

- The thicker the metal, the higher the amperage.
- The more coordinated you get with the foot pedal, the higher you can leave the amperage.
- Some conventional current ratios are: 1.6mm, 30 to 120 amps; 2.4mm, 80 to 240 amps; 3.2mm, 200 to 380 amps.

Method 2. Welding your Metal

1. Clean your welding material. Your surface must be clear of debris before you begin to weld.

- To prepare carbon steel, use a grinder or sander and polish it down to a bare, shiny metal.
- For aluminum, it is best to use a dedicated stainless steel wire brush.
- For stainless steel, just wipe down the weld area with some solvent on a rag. Make sure to store the rag and chemicals in a safe place before welding.

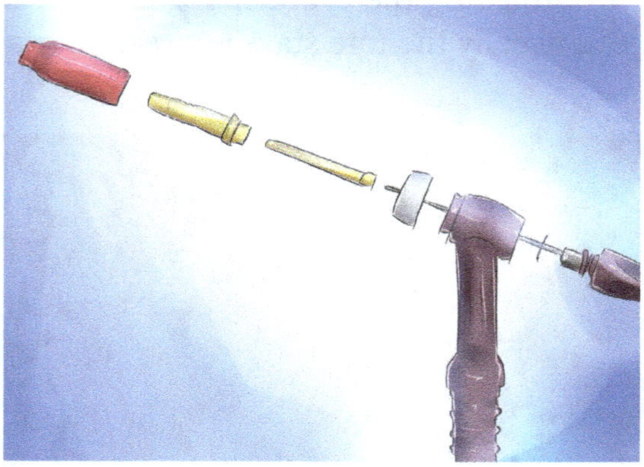

2. Insert the tungsten electrode into its collet. Unscrew the back of the electrode holder on the collet, insert the tungsten electrode, and screw the back on again. Generally, the electrode should hang about 1/4-inch away from the protective sheath on the collet.

3. Clamp the parts together. Use an angle iron and/or a flat bar with c-clamps to secure the parts you want to weld together.

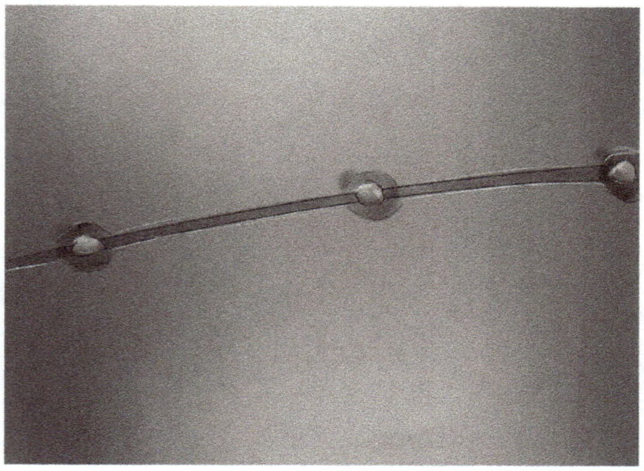

4. Tack weld the parts together. A tack weld is a very small weld that is intended to hold a part

in place until the final weld can be completed. Place tack welds every few inches where your two metals meet.

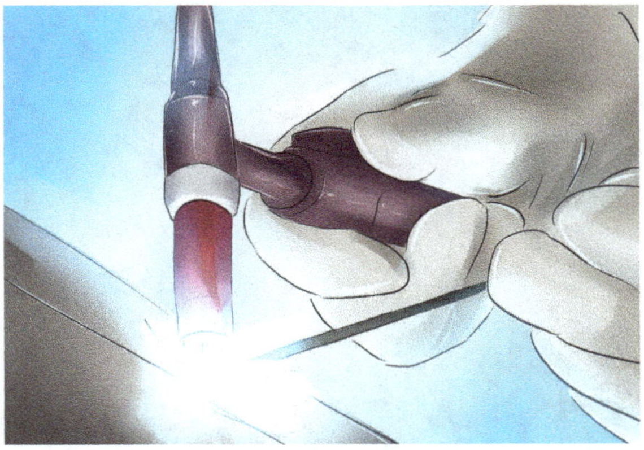

5. Hold the TIG torch in your hand. Be sure to hold it at about a 75 degree angle with the tungsten raised no more than 1/4-inch off the metal.

- Don't let the tungsten touch the work piece or it will contaminate your material.

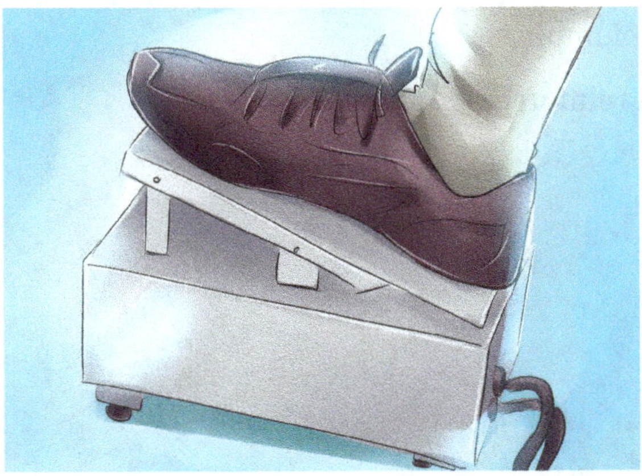

6. Practice using the foot pedals to control the heat. Your weld puddle should be about 1/4-inch wide. It is important to keep your puddle size consistent throughout the weld to avoid a messy finish.

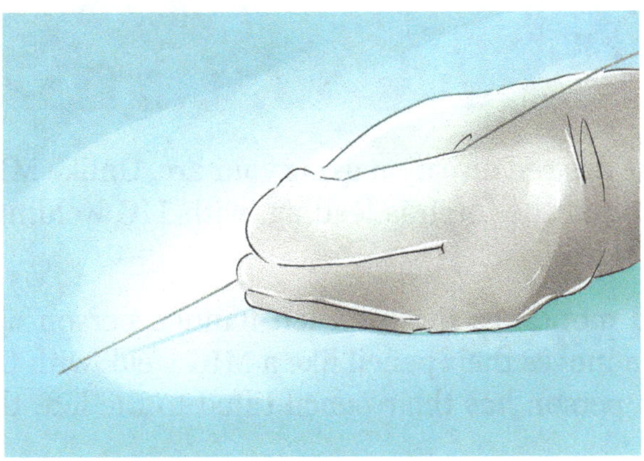

7. Pick up the filler rod in your other hand. Hold it so it rests horizontally at a 15 degree angle from the work piece at the base where the torch will heat the piece.

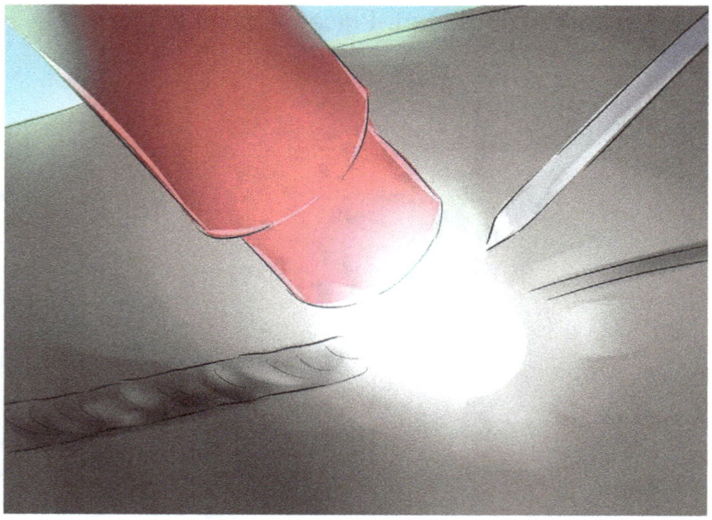

8. Use your torch to heat up the base metal. The heat of the arc will create a puddle, a pool of molten metal that is used to fuse the two pieces of metal together.

- Once there is a puddle on both pieces of metal, tap the filler rod into the molten puddle in quick dabs to avoid clumping.
- The filler rod adds a reinforcement layer for your weld.

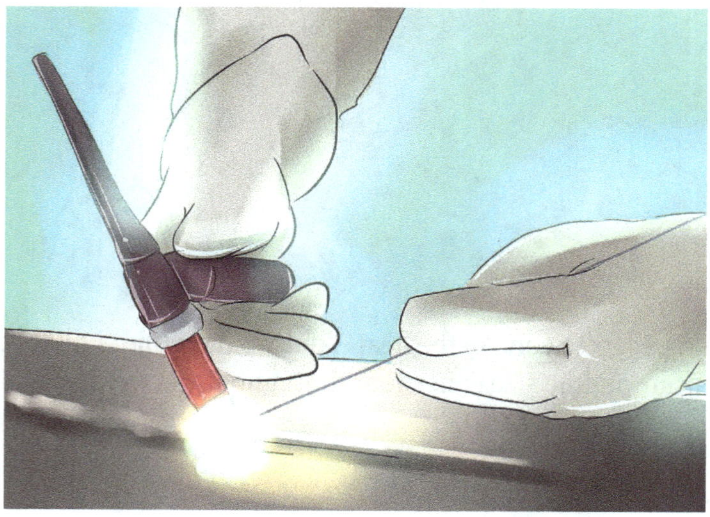

9. Advance the puddle in the desired direction using your arc. Unlike MIG welding, where you lead the puddle in the direction that the torch is leading, with TIG welding you push the puddle the opposite direction that the torch leans.

- Think of your hand motion as that of a left-handed person operating a pencil. While a right-handed person moves their pencil like a MIG weld, with the angles both tilted to the right, a left-handed person has their pencil tilted to the left, though they must push the pencil to the right.
- Continue to advance the puddle until you have welded the entire area desired and you've completed a TIG weld.

Method 3. Learning Different Types of Welds

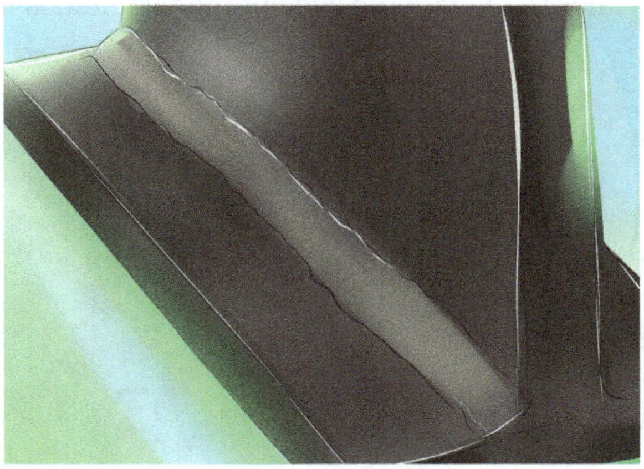

1. Fashion an easy fillet weld. Start off with a fillet weld to get the hang of TIG welding. A fillet weld is comprised of two metals joined at right angles. Run a weld puddle at a 45 degree angle to a 90 degree corner. A fillet weld should look like a triangle from the side.

2. Weld a lap joint. Form the weld puddle between the edge of an overlapping piece of metal and the surface of the bottom piece of metal. When these pieces fuse together, dip the filler rod into the puddle.

3. Make a T-joint to connect two pieces of metal at a right angle. Angle the torch so that there is

direct heat on the flat surface of the metal. Hold a shorter arc by extending the electrode beyond the ceramic cone. Place the filler rod where the edge of the two metals meet.

4. Melt a corner joint. Melt both edges of the metal where they meet in a point. Keep the weld puddle in the center of the joint where the two metals meet. You will need a significant amount of filler rod for a corner joint because the metals do not overlap.

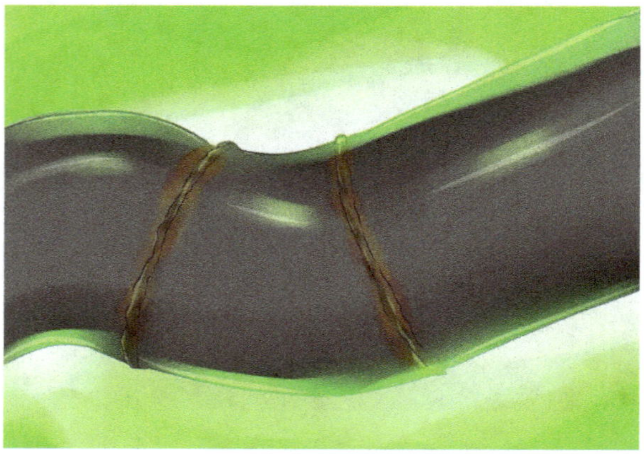

5. Create a butt weld. Center the weld pool on the adjoining edges of two pieces of metal. This requires more skill that the other types of weld because the metals do not overlap. When finishing, decrease the amperage to fill in the crater that forms.

Plasma Arc Welding

The the Plasma Arc Welding (PAW) and cutting process was invented by Robert M.Gage in 1953 and patented in 1957. This process was unique because it could do precision cutting on both thick and thin metal. It is also capable of spray coating hardening metal on other metals.

Principle

PAW is process where a coalescence is produced by the heat which is developed from a special setup between a tungsten alloy electrode and the water cooled nozzle (Non transferred ARC) or between a tungsten alloy electrode and the job (transferred ARC) . This process employs two different gases for two different purposes:

- One gas is used to form the Arc plasma.
- Second gas is used to shield the arc plasma.

Equipment:

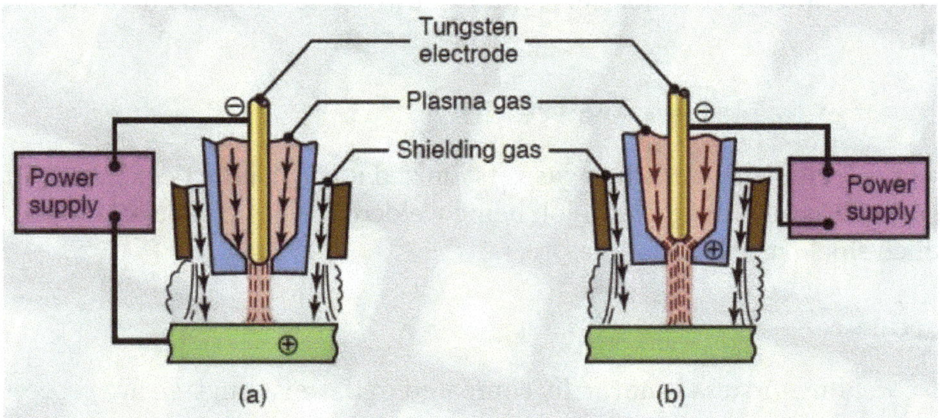

Power Source

The plasma arc is normally operated with a DC, drooping characteristic power source. Because its unique operating features are derived from the special torch arrangement and separate plasma and shielding gas flows, a plasma control console can be added on to a conventional TIG power source. Purpose-built plasma systems are also available. The plasma arc is not readily stabilised with sine wave AC. Arc reignition is difficult when there is a long electrode to workpiece distance and the plasma is constricted, Moreover, excessive heating of the electrode during the positive half-cycle causes balling of the tip which can disturb arc stability.

Special-purpose switched DC power sources are available. By imbalancing the waveform to reduce the duration of electrode positive polarity, the electrode is kept sufficiently cool to maintain a pointed tip and achieve arc stability.

Arc Starting

Although the arc is initiated using HF, it is first formed between the electrode and plasma nozzle. This 'pilot' arc is held within the body of the torch until required for welding then it is transferred to the workpiece. The pilot arc system ensures reliable arc starting and, as the pilot arc is maintained between welds, it obviates the need for HF which may cause electrical interference.

Electrode

The electrode used for the plasma process is tungsten-2%thoria and the plasma nozzle is copper. The electrode tip diameter is not as critical as for TIG and should be maintained at around 30-60 degrees. The plasma nozzle bore diameter is critical and too small a bore diameter for the current level and plasma gas flow rate will lead to excessive nozzle erosion or even melting. It is prudent to use the largest bore diameter for the operating current level.

Plasma and Shielding Gases

The normal combination of gases is argon for the plasma gas, with argon plus 2 to 5% hydrogen for the shielding gas. Helium can be used for plasma gas but because it is hotter this reduces the current rating of the nozzle. Helium's lower mass can also make the keyhole mode more difficult.

Working Concept of Plasma Arc Welding

The process works by ionizing gas.when gas gets ionized it can conduct electricity.The gas is used to transfer an electric arc to the piece of job being welded.The gas can be argon plus a secondary gas helium which shield arc weld puddle.

Working Process

The plasma arc welding process is normally compared to the gas tungsten arc process :

1. The plasma torch contains an electrode made out of tungsten fixed in a nozzle which is made of copper . The arc is started between the electrode and the tip of nozzle. Then the arc or flame is transferred to the material to be welded.
2. The small opening forces the gas to travel through a constricted opening or orifice. This concentrates the heat to smaller area. This ability allows welder to produce a very high quality weld.
3. The result is a process that gives higher welds speed ,less distortion, more consistent welds, less spatter and more control on the weld area.

Plasma welding process has two different modes that is :

1. Non transferred arc mode: In Non transferred arc mode the current flow is from the electrode inside the torch to the nozzle containing the orifice and back to the power supply. It is used for plasma spraying.
2. Transferred arc mode: In transferred arc mode the current is transferred from the tungsten electrode inside the welding torch through the orifice to the workpiece and back to power supply. It is used in welding metals.

Features

P	Protected electrode, offers long times before electrode maintenance (usually one 8 Hr Shift).
L	Low amperage welding capability (as low as 0.05 amp).
A	Arc consistency and gentle arc starting produce consistent welds, time after time.
S	Stable arc in arc starting and low amperage welding.
M	Minimal high frequency noise issues, HF only in pilot arc start, not for each weld.
A	Arc energy density reaches 3 times that of GTAW. Higher weld speeds possible.

W	Weld times as short as 5 msecs (.005 secs).
E	Energy density reduces heat affected zone, improves weld quality.
L	Length of arc benefit due to arc shape and even heat distribution.
D	Diameter of arc chosen via nozzle orifice.

Benefits

The full list of reasons for using the plasma welding process is lengthy but can be summarized into three main features where customers desire the advantages of at least one feature.

- Precision: The plasma process is generally more precise than conventional Tig (remember that enhanced power supplies can create an arc that is different to a conventional Tig arc) Plasma offers the following advantages over conventional Tig:
    - Stable, concentrated arc.
    - Forgiveness in arc length variations (Tig +/- 5%, Plasma +/- 15%).
- Small Part Welding:
    - Low amperage capability (many plasma power supplies go down to .1 amps).
    - Stable at low amps.
    - Gentle arc transfer (arc start) with no high frequency noise.
    - Short weld times possible (for spot welds - guidewires, tubes etc.)
- High Production Welding:
    - Long electrode life offers many more hours of welding than Tig before electrode contamination occurs.

In many applications, many of the unique advantages of plasma combine to benefit the overall welding process.

Applications

Small Part Welding: The plasma process can gently yet consistently start an arc to the tip of wires or other small components and make repeatable welds with very short weld time periods. This is advantageous when welding components such as needles, wires, light bulb filaments, thermocouples, probes and some surgical instruments.

Sealed Components: Medical and electronic components are often hermetically sealed via welding. The plasma process provides the ability to:

1. Reduce the heat input to the part
2. Weld near delicate insulating seals
3. Start the arc without high frequency electrical noise which could be damaging to the electrical internals.

Applications include Pressure and Electrical Sensors, Bellows, Seals, Cans, Enclosures, Microswitches, Valves, Electronic Components, Motors, Batteries, Miniature Tube to Fitting/Flange, Food and Dairy Equipment.

Tool Die & Mold Repair: A whole repair industry has sprung up to assist companies wishing to reuse components with slight nicks and dents from misuse or wear. The ability of modern micro-arc power supplies to gently start a low amperage arc and make repairs has provided users with a unique alternative to conventional repair and heat treatment. Both the Micro-Tig and micro-plasma welding processes are used for tool, die and mold repair. For outside edges the Plasma process offers great arc stability and requires less skill to control the weld puddle. To reach inside corners and crevices the TIG process allows the tungsten welding electrode to be extended in order to improve access.

Strip Metal Welding: The plasma process provides the ability to consistently transfer the arc to the workpiece and weld up to the edges of the weld joint. In automatic applications no Arc Distance Control is necessary for long welds and the process requires less maintenance to the torch components. This is especially advantageous in high volume applications where the material outgases or has surface contaminants.

Tube Mill Welding: Tube mills produce tube and pipe by taking a continuous strip of material and rollforming the edges upwards until the edges of the strip meet together at a weld station. At this point the welding process melts and fuses the edges of the tube together and the material exits the weld station as welded tube.

The output of the tube mill depends on the arc welding speed and total time spent welding. Each time the mill shuts down and starts up again there is a certain amount of scrap produced. Thus the most important issues to the tube mill user are:

1. Maximum tube mill weld speed obtainable.
2. Arc stability for optimum weld quality and consistency.
3. Maximum number of hours of welding electrode tip life.

Some tube mills employ plasma welding in order to get a combination of increased weld speed, improved weld penetration and maximum electrode life.

Submerged Arc Welding

Submerged arc welding is a process in which the joining of metals is produced by heating with an arc or arcs between a bare metal electrode or electrodes and the work.

Equipment

Equipment consists of a welding machine or power source, the wire feeder and control system, the welding torch for automatic welding or the welding gun and cable assembly for semiautomatic welding, the flux hopper and feeding mechanism, usually a flux recovery system, and a travel mechanism for automatic welding.

The power source for submerged arc welding must be rated for a 100 percent duty cycle, since the submerged arc welding operations are continuous and the length of time for making a weld may exceed 10 minutes.

If a 60 percent duty cycle power source is used, it must be derated according to the duty cycle curve for 100 percent operation.

When constant current is used, either ac or dc, the voltage sensing electrode wire feeder system must be used.

When constant voltage is used, the simpler fixed speed wire feeder system is used. The CV system is only used with direct current.

Both generator and transformer-rectifier power sources are used, but the rectifier machines are more popular.

Welding machines for submerged arc welding range in size from 300 amperes to 1500 amperes.

They may be connected in parallel to provide extra power for high-current applications.

Direct current power is used for semiautomatic applications, but alternating current power is used primarily with the machine or the automatic method.

Multiple electrode systems require specialized types of circuits, especially when ac is employed.

SAW Equipment

SAW or submerged arc welding equipment completing a weld. The weld starts on the right and moves left. The gray colored powder is flux.

For semiautomatic application, a welding gun and cable assembly are used to carry the electrode and current and to provide the flux at the arc.

A small flux hopper is attached to the end of the cable assembly.

The electrode wire is fed through the bottom of this flux hopper through a current pickup tip to the arc.

The flux is fed from the hopper to the welding area by means of gravity.

The amount of flux fed depends on how high the gun is held above the work.

The hopper gun may include a start switch to initiate the weld or it may utilize a “hot” electrode so that when the electrode is touched to the work, feeding will begin automatically.

For automatic welding, the torch is attached to the wire feed motor and includes current pickup tips for transmitting the welding current to the electrode wire.

The flux hopper is normally attached to the torch, and may have magnetically operated valves which can be opened or closed by the control system.

Other pieces of equipment sometimes used may include a travel carriage, which can be a simple tractor or a complex moving specialized fixture. A flux recovery unit is normally provided to collect the unused submerged arc flux and return it to the supply hopper.

Submerged arc welding system can become quite complex by incorporating additional devices such as seam followers, weavers, and work rovers.

SAW Welding Diagram

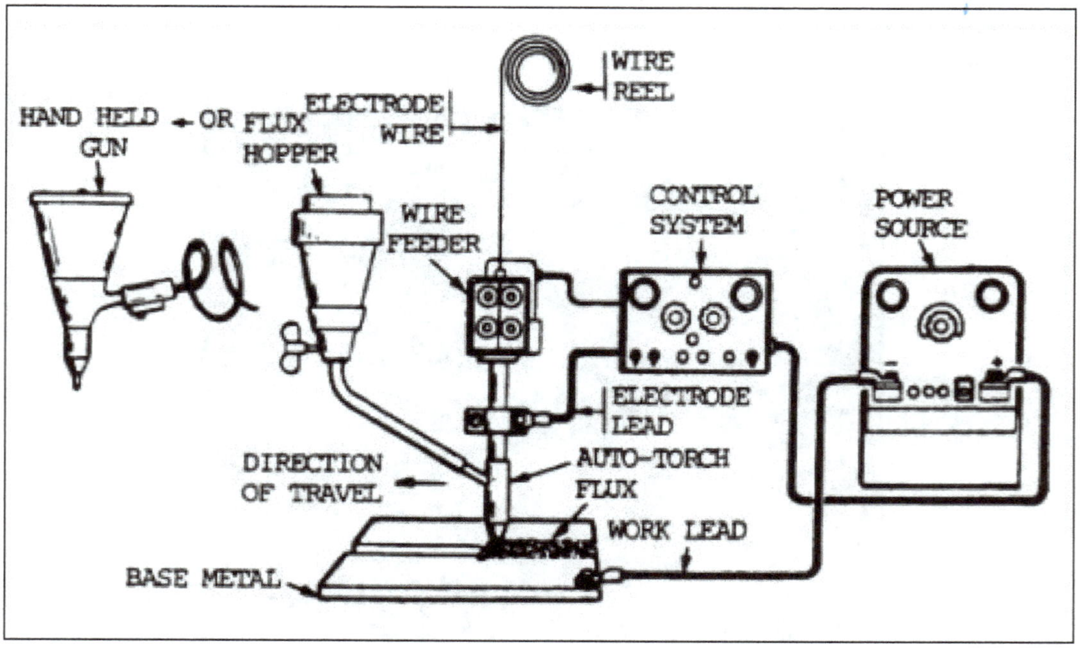

Block diagram of SAW (submerged arc welding) Equipment.

Advantages

The major advantages of the SAW or submerged arc welding process are:

1. High quality metal weld.
2. Extremely high speed and deposition rate
3. Smooth, uniform finished weld with no spatter.
4. Little or no smoke.
5. No arc flash, thus minimal need for protective clothing.
6. High utilization of electrode wire.
7. Easy automation for high-operator factor.
8. Normally, no involvement of manipulative skills.

SAW Pipe Welding

SAW welding process to build long steel piles to support ocean platform.

Major Uses

The submerged arc process is widely used in heavy steel plate fabrication work. This includes the welding of structural shapes, the longitudinal seam of larger diameter pipe, the manufacture of machine components for all types of heavy industry, and the manufacture of vessels and tanks for pressure and storage use. It is widely used in the shipbuilding industry for splicing and fabricating sub-assemblies, and by many other industries where steels are used in medium to heavy thicknesses. It is also used for surfacing and buildup work, maintenance, and repair.

SAW Welding

In SAW welding the flux and wire are separate. Both impact the properties of the weld, requiring the selection of the optimal combination by the engineer for each project.

Process Limitations

A major limitation of SAW (submerged arc welding) is its limitation of welding positions. The other limitation is that it is primarily used only to weld mild and low-alloy high-strength steels.

The high-heat input and slow-cooling cycle can be a problem when welding quenched and tempered steels. The heat input limitation of the steel in question must be strictly adhered to when using submerged arc welding. This may require the making of multipass welds where a single pass weld would be acceptable in mild steel. In some cases, the economic advantages may be reduced to the point where flux-cored arc welding or some other process should be considered.

In semiautomatic submerged arc welding, the inability to see the arc and puddle can be a disadvantage in reaching the root of a groove weld and properly filling or sizing.

Submerged Arc Welding

Demonstration of Saw Welding Process.

Principles of Operation

Process

The submerged arc welding process is shown by figure below. It utilizes the heat of an arc between a continuously fed electrode and the work.

Submerged Arc Welding Process Diagram

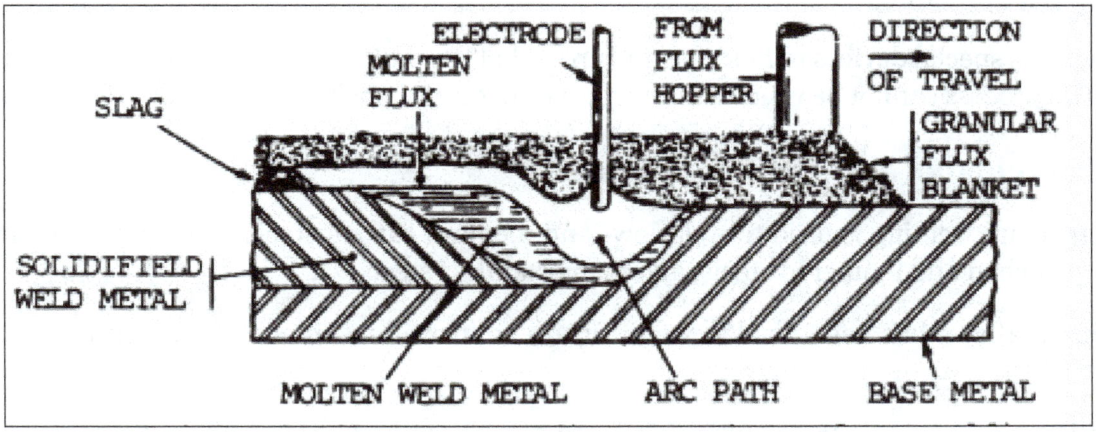

Process Diagram for SAW (submerged arc welding)

The heat of the arc melts the surface of the base metal and the end of the electrode. The metal melted off the electrode is transferred through the arc to the workpiece, where it becomes the deposited weld metal.

Shielding is obtained from a blanket of granular flux, which is laid directly over the weld area. The flux close to the arc melts and intermixes with the molten weld metal, helping to purify and fortify it.

The flux forms a glass-like slag that is lighter in weight than the deposited weld metal and floats on the surface as a protective cover.

The weld is submerged under this layer of flux and slag, hence the name submerged arc welding. The flux and slag normally cover the arc so that it is not visible.

The unmelted portion of the flux can be reused. The electrode is fed into the arc automatically from a coil. The arc is maintained automatically.

Travel can be manual or by machine. The arc is initiated by a fuse type start or by a reversing or retrack system.

Normal Method of Application and Position Capabilities

The most popular method of SAW application is the machine method, where the operator monitors the welding operation.

Second in popularity is the automatic method, where welding is a pushbutton operation. The process can be applied semiautomatically; however, this method of application is not too popular.

The process cannot be applied manually because it is impossible for a welder to control an arc that is not visible. The submerged arc welding process is a limited-position welding process.

The welding positions are limited because the large pool of molten metal and the slag are very fluid and will tend to run out of the joint. Welding can be done in the flat position and in the horizontal fillet position with ease.

Under special controlled procedures, it is possible to weld in the horizontal position, sometimes called 3 o'clock welding.

This requires special devices to hold the flux up so that the molten slag and weld metal cannot run away. The process cannot be used in the vertical or overhead position.

Metals Weldable and Thickness Range

Submerged arc welding is used to weld low- and medium-carbon steels, low-alloy high-strength steels, quenched and tempered steels, and many stainless steels.

Experimentally, it has been used to weld certain copper alloys, nickel alloys, and even uranium.

Metal thicknesses from 1/16 to 1/2 in. (1.6 to 12.7 mm) can be welded with no edge preparation. With edge preparation, welds can be made with a single pass on material from 1/4 to 1 in. (6.4 to 25.4 mm).

When multipass technique is used, the maximum thickness is practically unlimited. This information is summarized in table 10-22. Horizontal fillet welds can be made up to 3/8 in. (9.5 mm) in a single pass and in the flat position, fillet welds can be made up to 1 in. (25 mm) size.

Joint Design

Although the submerged arc welding process can utilize the same joint design details as the shielded metal arc welding process, different joint details are suggested for maximum utilization and efficiency of submerged arc welding. For groove welds, the square groove design can be used up to 5/8 in. (16 mm) thickness.

Beyond this thickness, bevels are required. Open roots are used but backing bars are necessary since the molten metal will run through the joint.

When welding thicker metal, if a sufficiently large root face is used, the backing bar may be eliminate. However, to assure full penetration when welding from one side, backing bars are recommended. Where both sides are accessible, a backing weld can be made which will fuse into the original weld to provide full penetration.

Welding Circuit and Current

The SAW or submerged arc welding process uses either direct or alternating current for welding power. Direct current is used for most applications which use a single arc. Both direct current electrode positive (DCEP) and electrode negative (DCEN) are used.

The constant voltage type of direct current power is more popular for submerged arc welding with 1/8 in. (3.2 mm) and smaller diameter electrode wires.

The constant current power system is normally used for welding with 5/3 2 in. (4 mm) and larger-diameter electrode wires. The control circuit for CC power is more complex since it attempts to duplicate the actions of the welder to retain a specific arc length. The wire feed system must sense the voltage across the arc and feed the electrode wire into the arc to maintain this voltage. As conditions change, the wire feed must slow down or speed up to maintain the prefixed voltage across the arc. This adds complexity to the control system. The system cannot react instantaneously. Arc starting is more complicated with the constant current system since it requires the use of a reversing system to strike the arc, retract, and then maintain the preset arc voltage.

For SAW ac welding, the constant current power is always used. When multiple electrode wire systems are used with both ac and dc arcs, the constant current power system is utilized. The constant voltage system, however, can be applied when two wires are fed into the arc supplied by a single power source. Welding current for submerged arc welding can vary from as low as 50 amperes to as high as 2000 amperes. Most submerged arc welding is done in the range of 200 to 1200 amperes.

Deposition Rates and Weld Quality

The deposition rates of the submerged arc welding process are higher than any other arc welding process. There are at least four related factors that control the deposition rate of submerged arc welding: polarity, long stickout, additives in the flux, and additional electrodes. The deposition rate is the highest for direct current electrode negative (DCEN). The deposition rate for alternating current is between DCEP and DCEN. The polarity of maximum heat is the negative pole.

The deposition rate with any welding current can be increased by extending the "stickout." This is the distance from the point where current is introduced into the electrode to the arc. When using "long stickout" the amount of penetration is reduced. The deposition rates can be increased by metal additives in the submerged arc flux. Additional electrodes can be used to increase the overall deposition rate.

The quality of the weld metal deposited by the submerged arc welding process is high. The weld metal strength and ductility exceeds that of the mild steel or low-alloy base material when the correct combination of electrode wire and submerged arc flux is used. When submerged arc welds are made by machine or automatically, the human factor inherent to the manual welding processes is eliminated. The weld will be more uniform and free from inconsistencies. In general, the weld bead size per pass is much greater with submerged arc welding than with any of the other arc welding processes. The heat input is higher and cooling rates are slower. For this reason, gases are allowed more time to escape. Additionally, since the submerged arc slag is lower in density than the weld metal, it will float out to the top of the weld. Uniformity and consistency are advantages of this process when applied automatically.

Several problems may occur when using the semiautomatic application method. The electrode wire may be curved when it leaves the nozzle of the welding gun. This curvature can cause the arc to be struck in a location not expected by the welder. When welding in fairly deep grooves, the curvature may cause the arc to be against one side of the weld joint rather than at the root. This will cause incomplete root fusion. Flux will be trapped at the root of the weld. Another problem with semiautomatic welding is that of completely filling the weld groove or maintaining exact size, since the weld is hidden and cannot be observed while it is being made. This requires making an extra pass. In some cases, too much weld is deposited. Variations in root opening affect the travel speed. If travel speed is uniform, the weld may be under- or overfilled in different areas. High operator skill will overcome this problem.

There is another quality problem associated with extremely large single-pass weld deposits. When these large welds solidify, the impurities in the melted base metal and in the weld metal all collect at the last point to freeze, which is the centerline of the weld. If there is sufficient restraint and enough impurities are collected at this point, centerline cracking may occur. This can happen when making large single-pass flat fillet welds if the base metal plates are 45° from flat. A simple solution is to avoid placing the parts at a true 45° angle. It should be varied approximately 10° so that the root of the joint is not in line with the centerline of the fillet weld. Another solution is to make multiple passes rather than attempting to make a large weld in a single pass.

Another quality problem has to do with the hardness of the deposited weld metal. Excessively hard weld deposits contribute to cracking of the weld during fabrication or during service. A maximum hardness level of 225 Brinell is recommended. The reason for the hard weld in carbon and low-alloy steels is too rapid cooling, inadequate postweld treatment, or excessive alloy pickup in the weld metal. Excessive alloy pickup is due to selecting an electrode that has too much alloy, selecting a flux that introduces too much alloy into the weld, or the use of excessively high welding voltages.

In automatic and machine welding, defects may occur at the start or at the end of the weld. The best solution is to use runout tabs so that starts and stops will be on the tabs rather than on the product.

Weld Schedules

The submerged arc welding process applied by machine or fully automatically should be done in accordance with welding procedure schedules. All of the welds made by this procedure should pass qualification, tests, assuming that the correct electrode and flux have been selected. If the schedules are varied more than 10 percent, qualification tests should be performed to determine the weld quality.

Welding Variables

The welding variables for submerged arc welding are similar to the other arc welding processes, with several exceptions.

In submerged arc welding, the electrode type and the flux type are usually based on the mechanical properties required by the weld. The electrode size is related to the weld joint size and the current recommended for the particular joint. This must also be considered in determining the number of passes or beads for a particular joint. Welds for the same joint dimension can be made in many or few passes, depending on the weld metal metallurgy desired. Multiple passes usually deposit higher-quality weld metal. Polarity is established initially and is based on whether maximum penetration or maximum deposition rate is required.

The major variables that affect the weld involve heat input and include the welding current, arc voltage, and travel speed. Welding current is the most important. For single-pass welds, the current should be sufficient for the desired penetration without burn-through. The higher the current, the deeper the penetration. In multi-pass work, the current should be suitable to produce the size of the weld expected in each pass. The welding current should be selected based on the electrode size. The higher the welding current, the greater the melt-off rate (deposition rate).

The arc voltage is varied within narrower limits than welding current. It has an influence on the bead width and shape. Higher voltages will cause the bead to be wider and flatter. Extremely high arc voltage should be avoided, since it can cause cracking. This is because an abnormal amount of flux is melted and excess deoxidizers may be transferred to the weld deposit, lowering its ductility. Higher arc voltage also increases the amount of flux consumed. The low arc voltage produces a stiffer arc that improves penetration, particularly in the bottom of deep grooves. If the voltage is too low, a very narrow bead will result. It will have a high crown and the slag will be difficult to remove.

Travel speed influences both bead width and penetration. Faster travel speeds produce narrower beads that have less penetration. This can be an advantage for sheet metal welding where small beads and minimum penetration are required. If speeds are too fast, however, there is a tendency for undercut and porosity, since the weld freezes quicker. If the travel speed is too slow, the electrode stays in the weld puddle too long. This creates poor bead shape and may cause excessive spatter and flash through the layer of flux.

The secondary variables include the angle of the electrode to the work, the angle of the work itself, the thickness of the flux layer, and the distance between the current pickup tip and the arc. This latter factor, called electrode "stickout," has a considerable effect on the weld. Normally, the distance between the contact tip and the work is 1 to 1-1/2 in. (25 to 38 mm). If the stickout is increased beyond this amount, it will cause preheating of the electrode wire, which will greatly increase the deposition rate. As stickout increases, the penetration into the base metal decreases. This factor must be given serious consideration because in some situations the penetration is required.

The depth of the flux layer must also be considered. If it is too thin, there will be too much arcing through the flux or arc flash. This also may cause porosity. If the flux depth is too heavy, the weld may be narrow and humped. Too many small particles in the flux can cause surface pitting since the gases generated in the weld may not be allowed to escape. These are sometimes called peck marks on the bead surface.

Tips for using the Process

One of the major applications for submerged arc welding is on circular welds where the parts are rotated under a fixed head. These welds can be made on the inside or outside diameter. Submerged arc welding produces a large molten weld puddle and molten slag which tends to run. This dictates that on outside diameters, the electrode should be positioned ahead of the extreme top, or 12 o'clock position, so that the weld metal will begin to solidify before it starts the downside slope. This becomes more of a problem as the diameter of the part being welded gets smaller. Improper electrode position will increase the possibility of slag entrapment or a poor weld surface. The angle of the electrode should also be changed and pointed in the direction of travel of the rotating part.

When the welding is done on the inside circumference, the electrode should be angled so that it is ahead of bottom center, or the 6 o'clock position.

Sometimes the work being welded is sloped downhill or uphill to provide different types of weld bead contours. If the work is sloped downhill, the bead will have less penetration and will be wider. If the weld is sloped uphill, the bead will have deeper penetration and will be narrower. This is based on all other factors remaining the same.

The weld will be different depending on the angle of the electrode with respect to the work when the work is level. This is the travel angle, which can be a drag or push angle. It has a definite effect on the bead contour and weld metal penetration.

One side welding with complete root penetration can be obtained with submerged arc welding. When the weld joint is designed with a tight root opening and a fairly large root face, high current and electrode positive should be used. If the joint is designed with a root opening and a minimum root face, it is necessary to use a backing bar, since there is nothing to support the molten weld metal. The molten flux is very fluid and will run through narrow openings. If this happens, the weld metal will follow and the weld will burn through the joint. Backing bars are needed whenever there is a root opening and a minimum root face.

Copper backing bars are useful when welding thin steel. Without backing bars, the weld would tend to melt through and the weld metal would fall away from the joint. The backing bar holds the weld metal in place until it solidifies. The copper backing bars may be water cooled to avoid the possibility of melting and copper pickup in the weld metal. For thicker materials, the backing may be submerged arc flux or other specialized type flux.

Variations of the SAW Process

There are a large number of variations to the process that give submerged arc welding additional capabilities. Some of the more popular variations are:

1. Two-wire systems--same power source.
2. Two-wire systems--separate power source.
3. Three-wire systems--separate power source.
4. Strip electrode for surfacing.
5. Iron powder additions to the flux.
6. Long stickout welding.
7. Electrically "cold" filler wire.

Multi-wire Systems

The multi-wire systems offer advantages since deposition rates and travel speeds can be improved by using more electrodes. There are two methods of utilizing two electrodes, one with a single-power source and one with two power sources. When a single-power source is used, the same drive rolls are used for feeding both electrodes into the weld. When two power sources are used, individual wire feeders must be used to provide electrical insulation between the two electrodes. With two electrodes and separate power, it is possible to utilize different polarities on the two electrodes or to utilize alternating current on one and direct current on the other. The electrodes can be placed side by side. This is called transverse electrode position. They can also be placed one in front of the other in the tandem electrode position.

Two-wire Tandem

The two-wire tandem electrode position with individual power sources is used where extreme penetration is required. The leading electrode is positive with the trailing electrode negative. The first electrode creates a digging action and the second electrode fills the weld joint. When two dc arcs are in close proximity, there is a tendency for arc interference between them. In some cases, the second electrode is connected to alternating current to avoid the interaction of the arc.

Three-wire Tandem System

The three-wire tandem system normally uses ac power on all three electrodes connected to three-phase power systems. These systems are used for making high-speed longitudinal seams for large-diameter pipe and for fabricated beams. Extremely high currents can be used with correspondingly high travel speeds and deposition rates.

The Strip Welding System

The strip welding system is used to overlay mild and alloy steels usually with stainless steel. A wide bead is produced that has a uniform and minimum penetration. It is used for overlaying the inside of vessels to provide the corrosion resistance of stainless steel while utilizing the strength and economy of the low-alloy steels for the wall thickness. A strip electrode feeder is required and special flux is normally used. When the width of the strip is over 2 in. (51 mm), a magnetic arc oscillating device is used to provide for even burn-off of the strip and uniform penetration.

Other Options

Another way of increasing the deposition rate of submerged arc welding is to add iron base ingredients to the joint under the flux. The iron in this material will melt in the heat of the arc and will become part of the deposited weld metal. This increases deposition rates without decreasing weld metal properties. Metal additives can also be used for special surfacing applications. This variation can be used with single-wire or multi-wire installations.

Another variation is the use of an electrically "cold" filler wire fed into the arc area. The "cold" filler rod can be solid or flux-cored to add special alloys to the weld metal. By regulating the addition of the proper material, the properties of the deposited weld metal can be improved. It is possible to utilize a flux-cored wire for the electrode, or for one of the multiple electrodes to introduce special alloys into the weld metal deposit. Each of these variations requires special engineering to ensure that the proper material is added to provide the desired deposit properties.

Typical Applications

The submerged arc welding process is widely used in the manufacture of most heavy steel products. These include pressure vessels, boilers, tanks, nuclear reactors, chemical vessels, etc. Another use is in the fabrication of trusses and beams. It is used for welding flanges to the web. The heavy equipment industry is a major user of submerged arc welding.

Materials Used

Two materials are used in submerged arc welding: the welding flux and the consumable electrode wire.

Submerged arc welding flux shields the arc and the molten weld metal from the harmful effects of atmospheric oxygen and nitrogen. The flux contains deoxidizers and scavengers which help to remove impurities from the molten weld metal. Flux also provides a means of introducing alloys into the weld metal. As this molten flux cools to a glassy slag, it forms a covering which protects the surface of the weld. The unmelted portion of the flux does not change its form and its properties are not affected, so it can be recovered and reused. The flux that does melt and forms the slag covering must be removed from the weld bead. This is easily done after the weld has cooled. In many cases, the slag will actually peel without requiring special effort for removal. In groove welds, the solidified slag may have to be removed by a chipping hammer.

Fluxes are designed for specific applications and for specific types of weld deposits. Submerged arc fluxes come in different particle sizes. Many fluxes are not marked for size of particles because the size is designed and produced for the intended application.

Carbon Arc Welding

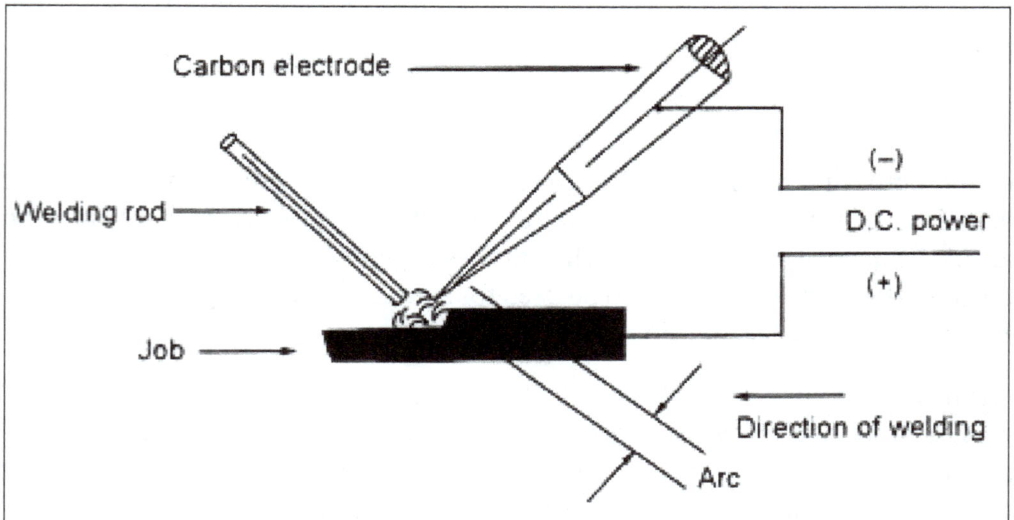

Carbon Arc Welding

Carbon Arc Welding (CAW) is a welding process, in which heat is generated by an electric arc struck between an carbon electrode and the work piece. The arc heats and melts the work pieces edges, forming a joint. Carbon arc welding is the oldest welding process.

In this method an electric arc is produced between the carbon electrode and the 'work'. A rod of carbon is used as negative (-) pole and the 'work' being welded as positive (+) pole. The carbon electrode does not melt itself. It is a non-consumable electrode.

A filler rod is used separately at the welding joint. This process is used for welding both ferrous and non-ferrous metals like steel sheet, Brass, Bronze, Gun metal, M/S-sheet, carbon-steel etc. The temperature rises to approximately 600°C.

In this process any metal can be joined by Arc-Brazing. Carbon flame may produce this type of welding—"Twin-Carbon-Arc Brazing".

Procedure of Welding with Carbon Arc

The arc is established between the work-piece and a carbon electrode held in the electrode-holder. A long carbon arc is often desirable in order to prevent contamination of the weld metal with the carbon monoxide given, off by the carbon electrode under the energy of the arc.

It is essential that the welding circuit be set-up for straight polarity. If a reversed-polarity setting is used, the arc will not be stable and a carbon residue will be left to become entrapped in the weld.

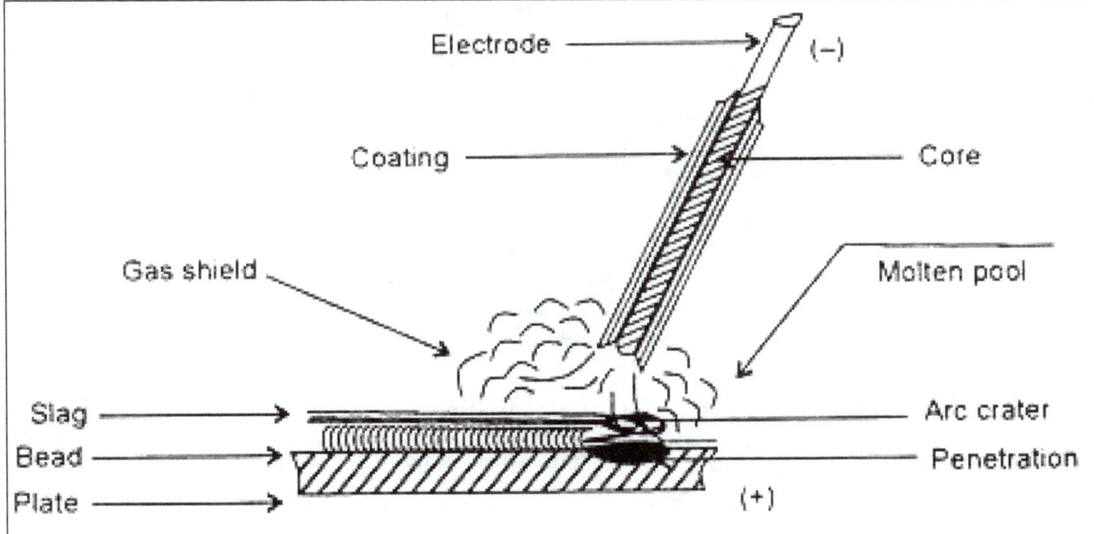

Manual Metal Arc Welding without Carbon Electrode

In the carbon arc welding of light-gauge plates, the use of filler rod may not be necessary. But in welding of heavy plates of filler rod—consisting of the same material as the base metal—it is generally used. In the latter procedure, a molten pool is first formed in the 'work'.

The filler rod is then inserted in the pool, the arc being directed against it just above the surface of the molten metal. The end of the rod is thus melted-off and deposited in the pool. The arc is played to and fro in the pool, thoroughly melting and mixing with the metal.

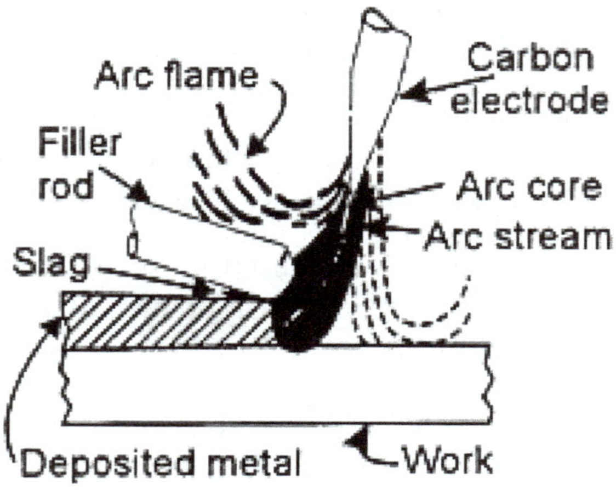

Welding with the Carbon Arc

No filler rod is necessary to make a lap weld on light-gauge metal with a carbon electrode. The electrode should be kept approximately perpendicular to the two plates and slightly over the edge of the top plate.

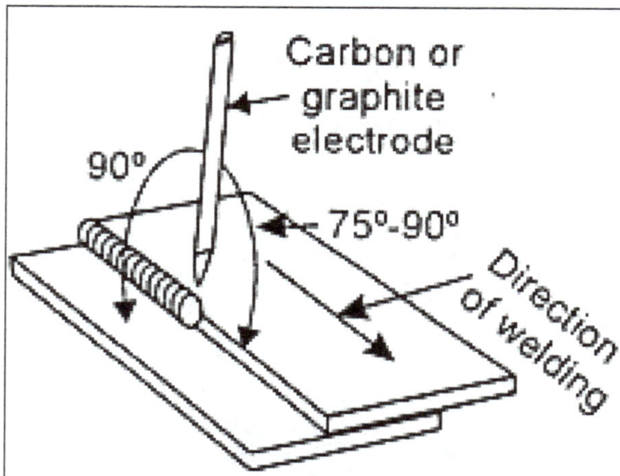

Making lap weld on light-gauge sheet metal with carbon electrode and no filler

The arc should fuse through the upper plate into the lower one. The amount of fusion will depend upon the speed with which the arc is moved over the surface. Slow to and fro movement increases penetration, but the fusing of a wide to and fro movement increases penetration. For less fusion, the current should be increased and the electrode advanced at a faster rate.

In welding a butt joint, a filler rod should be used— even for the lighter gauge metals. The electrode should be held almost perpendicular to the surface of the plates. The arc may be directed backwards into the weld, with the filler rod between the carbon electrode and the deposited weld.

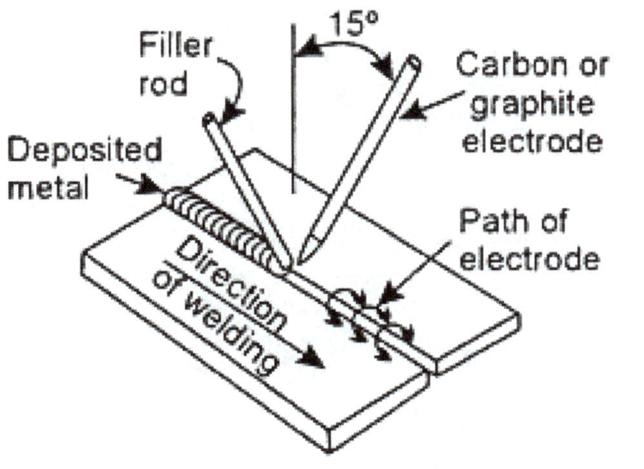

Making butt weld with carbon electrode and filler rod by using "backhand" technique

There are some techniques where the electrode is pointed away from the finished weld and the filler rod precedes the electrode in the direction of welding. The arc should be moved in a horseshoe or circular motion, and the edges of the plates should be fused by working along them for about 2-3 mm.

The heavier section plates should be butted and then tacked together with a space of 2-3 mm between the adjacent edges. These should be welded by using the filler-rod.

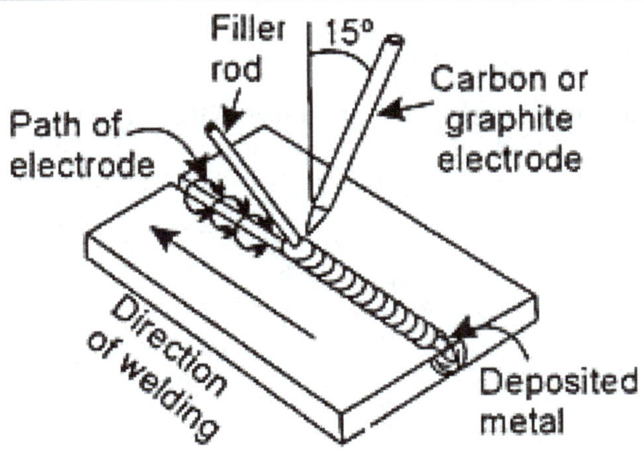

Making butt weld with carbon electrode and filler rod by using "forehand" technique

For edge welding light gauge sheets or plates they should be tacked together with the faces of the plates touching and with edges lined-up. The tacked assembly should be set-up with the edges on top. In making edge joint on light gauge sheet metal use carbon electrode and no filler rod.

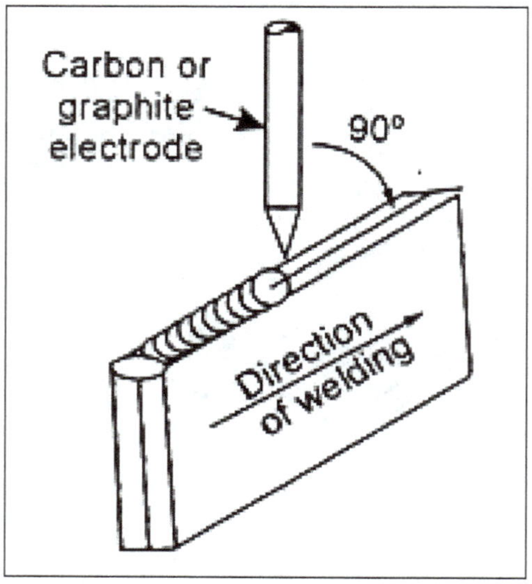

Making edge weld on light-gauge sheet metal with carbon electrode and no filler rod

Advantages of Carbon Arc Welding

- Low cost of equipment and welding operation;
- High level of operator skill is not required;
- The process is easily automated;
- Low distortion of work piece.

Disadvantages of Carbon Arc Welding

- Unstable quality of the weld (porosity);
- Carbon of electrode contaminates weld material with carbides.

Permissions

All chapters in this book are published with permission under the Creative Commons Attribution Share Alike License or equivalent. Every chapter published in this book has been scrutinized by our experts. Their significance has been extensively debated. The topics covered herein carry significant information for a comprehensive understanding. They may even be implemented as practical applications or may be referred to as a beginning point for further studies.

We would like to thank the editorial team for lending their expertise to make the book truly unique. They have played a crucial role in the development of this book. Without their invaluable contributions this book wouldn't have been possible. They have made vital efforts to compile up to date information on the varied aspects of this subject to make this book a valuable addition to the collection of many professionals and students.

This book was conceptualized with the vision of imparting up-to-date and integrated information in this field. To ensure the same, a matchless editorial board was set up. Every individual on the board went through rigorous rounds of assessment to prove their worth. After which they invested a large part of their time researching and compiling the most relevant data for our readers.

The editorial board has been involved in producing this book since its inception. They have spent rigorous hours researching and exploring the diverse topics which have resulted in the successful publishing of this book. They have passed on their knowledge of decades through this book. To expedite this challenging task, the publisher supported the team at every step. A small team of assistant editors was also appointed to further simplify the editing procedure and attain best results for the readers.

Apart from the editorial board, the designing team has also invested a significant amount of their time in understanding the subject and creating the most relevant covers. They scrutinized every image to scout for the most suitable representation of the subject and create an appropriate cover for the book.

The publishing team has been an ardent support to the editorial, designing and production team. Their endless efforts to recruit the best for this project, has resulted in the accomplishment of this book. They are a veteran in the field of academics and their pool of knowledge is as vast as their experience in printing. Their expertise and guidance has proved useful at every step. Their uncompromising quality standards have made this book an exceptional effort. Their encouragement from time to time has been an inspiration for everyone.

The publisher and the editorial board hope that this book will prove to be a valuable piece of knowledge for students, practitioners and scholars across the globe.

Index

www.ingramcontent.com/pod-product-compliance
Lightning Source LLC
LaVergne TN
LVHW080646210926
848485LV00003B/334